CHIRALITY IN BIOLOGICAL NANOSPACES

Reactions in Active Sites

CHIRALITY IN BIOLOGICAL NANOSPACES

Reactions in Active Sites

Nilashis Nandi

CRC Press
Taylor & Francis Group
Boca Raton London New York

CRC Press is an imprint of the
Taylor & Francis Group, an **informa** business

CRC Press
Taylor & Francis Group
6000 Broken Sound Parkway NW, Suite 300
Boca Raton, FL 33487-2742

First issued in paperback 2017

© 2012 by Taylor and Francis Group, LLC
CRC Press is an imprint of Taylor & Francis Group, an Informa business

No claim to original U.S. Government works

ISBN-13: 978-1-4398-4002-3 (hbk)
ISBN-13: 978-1-138-11269-8 (pbk)

Library of Congress Cataloging-in-Publication Data

Nandi, Nilashis.
    Chirality in biological nanospaces : reactions in active sites / Nilashis Nandi.
        p. ; cm.
    Includes bibliographical references and index.
    Summary: "Vital life processes occur within the active sites of large biological macromolecules such as proteins, nucleic acids, and lipids. These nanodimensional structures greatly accelerate biological reactions. The chirality of the reactants also has a strong influence on the process, but its importance in such biological reactions has only recently begun to be understood. This book explores the influence of chirality on reaction mechanisms in such biological nanospaces. The text addresses the influence of the chirality of amino acid and sugar in the active sites of transferase, oxidoreductases, hydrolases, lysases, isomerase, ligases, and other systems. It also covers ribosomal architecture"--Provided by publisher.
    Summary: "Discusses the relationship of molecular chirality and biology. Diverse natural biological molecules are dissymmetric and are discussed. Also described are the interactions between molecules in a confined state in biological systems and the resulting fascinating properties"--Provided by publisher.
    ISBN 978-1-4398-4002-3 (hardback : alk. paper)
    1. Chirality. 2. Biomolecules. I. Title.
    [DNLM: 1. Enzymes--chemistry. QU 135]

QP517.C57N36 2011
372'.33--dc22
                                        2011006151

Visit the Taylor & Francis Web site at
http://www.taylorandfrancis.com

and the CRC Press Web site at
http://www.crcpress.com

Dedication

Dedicated to my research students for sharing their interest in my research, and to my family members for their unstinting support and forbearance

Contents

Preface

The relationship of molecular chirality and biology is intriguing. Our curiosity prompts the question as to why diverse natural biological molecules are overwhelmingly dissymmetric. Is it so that nature uses the dissymmetry of biological molecules to create specificity in the interaction with other molecules? Does the myriad network of interaction aim to control the functionality of a bewildering variety of molecular machines with ease, fidelity, and efficiency? If so, how? Whatever the answer is, eventually the orientation-dependent interaction between molecules in the confined state in biological systems is responsible for the manifestation of many fascinating properties.

Enzymes exhibit remarkable capacities for discrimination as they carry out life processes. Despite the amazing variety of the structure and function of the enzymes, it is expected that unifying principles exist in their structural organization. Evolution must have developed, retained, and continued to further develop these minimal features for the better efficacy of a reaction. Active site structures of enzymes are nanosized spaces. Their confined geometries are designed by evolution to recognize and capture the specific substrates, place them in favored geometry, and finally drive forward the formation of the product to continue life processes. It is tempting to look into the active site structure and the discrimination therein to understand how nature utilizes the chiral structures of molecules in carrying out vital biological reactions much more efficiently than the corresponding synthetic processes. The conserved features of the organization of the active site structure of enzymes indicate that a network of electromagnetic interactions controls the remarkable fidelity of the reaction. The pattern of interaction is an outcome of the intricate interplay of electrostatic (such as charge–charge, charge–dipolar, polar–polar, induction type, and hydrogen bonding, to name a few), hydrophobic, or van der Waals interactions. The network of interaction depends on the orientation-dependent arrangement of the active site residues around the reactants, and confinement of the reactants is important here. The evolutionary processes develop the structure of the active site and network of interaction in order to perform reactions with improved speed, accuracy, and efficiency. With this perception in view, this book aims to provide the author's perspective about the influence of chirality in driving reactions within enzymatic cavities of nanodimension.

Chirality embraces numerous biological, biomimetic, and nonbiological systems, and the topic touches an extremely broad area of phenomena. Tremendous advances are being made in chirality-related fields using principles of physics, chemistry, biology, and mathematics. Often, chirality shows its manifestation in seemingly disparate phenomena. This book is not intended to provide a compendium of the wide variety of manifestations of chiralities in different systems, and no attempt is made to be exhaustive in this respect. Enzymes being the workhorses of the cells, their workings are of high significance and study in this field is greatly advancing. A vast body of literature exists about the chiral specificity of different enzymes. Since research on the stereoselectivity in enzymatic reactions is continuously growing, an anthology of stereoselectivity in enzymatic reactions is beyond the scope of this book. Consequently, if the reader is interested in learning about a specific class of enzyme, its structure, or biochemistry related to stereoselectivity, he or she is advised to look elsewhere for that information. On the other hand, readers will gain from this book a clear view of how the interactions between the active site residues and the substrate molecules influence reactions.

The author has written a brief review on the related topic in the *International Reviews in Physical Chemistry*, and his curiosity with the subject grew into this book. It was apparent during the time of writing the review, and still now, that there is a dearth of books that concentrate exclusively on the interaction of chiral molecules within the active site structure. This book evolved to fill this gap. It is intended primarily for graduate students, teachers, and researchers in the fields of chemistry, biology, medicine, pharmacy, and related interdisciplinary academia such as nanobiology and nanochemistry. One needs only basic ideas of intermolecular interaction as applicable to biophysical chemistry to make use of this book. The theoretical developments have been described at a simple, unsophisticated level throughout. As the subject is in a state of constant evolution and a comprehensive understanding is yet to be achieved in many arenas, constructive suggestions are most welcome.

It is our hope that the views presented in this book shall be further developed to design custom-made novel biocatalysts with better efficiency. Often a trial-and-error method is employed to develop synthetic enzymes, and the process is time consuming and expensive. A combination of crystallographic studies with electronic structure based computational analysis as described in this book may lead to future elucidation of new drugs that can target biological active sites with better efficacy.

Acknowledgments

It is a pleasure to thank research students Sindrila Dutta Banik and Krishnan Thirumoorthy for sharing the author's own interest in the chiral discrimination in biological systems. It is a delight to witness their dedication in conducting research work related to chiral discrimination in biological systems in the author's laboratory.

A large part of the research work described in this book was conducted at the University of Kalyani and Birla Institute of Technology and Science–Pilani. It is a pleasure to thank the enthusiastic students at these institutions for their active participation in the learning process. Their questions and responses stimulated the author's interest in biophysical chemistry as a constantly developing subject.

The author has cherished the delight of learning from all the teachers throughout his academic career and owes a great deal to their contributions in shaping his perspective on scientific principles. It is a pleasure to thank Dieter Vollhardt and Biman Bagchi for collaborations on various aspects of chirality in biomimetic systems.

The author is immensely thankful to Sindrila Dutta Banik for her assistance regarding many aspects of the preparation of this book—checking the manuscript and references, preparing most of the artwork, as well as the cover art, to name a few. Her help has been immeasurable.

It is a great pleasure to acknowledge the support of all the family members and friends. The patience of Anwesha Nandi and Shamita Goswami at home made any hardship during the preparation of the manuscript much easier to bear. This book could not have been completed without their indulgence in allowing the author to work unusual and long hours. The author takes this opportunity to express appreciation for the encouragement received from Shibshankar Nandi, Sunity Nandi, and the late Nirendra Kisore Nandi as well.

The Department of Science and Technology, Government of India supported the research work on chiral discrimination in enzymatic reactions carried out in the author's laboratory and thanks are extended to them. Appreciation is also due to the Council of Scientific and Industrial Research, Government of India and Alexander von Humboldt foundation,

Federal Republic of Germany for their support related to the research work in biomimetic systems in the author's laboratory.

Finally, it is a delight to thank Lance Wobus of Taylor & Francis Group who proposed and initiated the writing of this book and waited patiently. The author also thanks Taylor and Francis Group, CRC Press and especially David Fausel, Jennifer Derima, Judith Simon, Cristina Escalante, and Andrea Grant for their assistance in the production of this book.

About the author

Nilashis Nandi was born in Cooch Behar, West Bengal, India (1965). He received his B.Sc. (Hons.) (1983) and M.Sc. (1985) degrees from North Bengal University and Ph.D. (1992) from Visva Bharati University. He became a postdoctoral fellow at the Indian Institute of Science, India (1993–1997), a J.S.P.S. postdoctoral fellow at Nagoya University, Japan (1997–1999), and an Alexander von Humboldt postdoctoral fellow at the Max Planck Institute of Colloids and Interfaces, Germany (1999–2000). Dr. Nandi was a faculty member in the chemistry group of Birla Institute of Technology and Science, Pilani, India from 2001–2007 and became a professor in the Department of Chemistry, University of Kalyani in 2008 where he has worked ever since. His research interest is focused on theoretical and computational studies in biophysical chemistry.

List of abbreviations

aaRS: Aminoacyl tRNA synthetase
aatRNA: Aminoacyl transfer RNA
ADH: Alcohol oxidoreductase
AIDS: Acquired immune deficiency syndrome
AMP: Adenosine monophosphate
ATP: Adenosine triphosphate
BactnABC: Bacteroides thetaiotaomicron chondroitinase ABC
BAM: Brewster angle microscopy
BCH: 2'-deoxy-3'-thiacytidine
BS: *Bacillus stearothermophilus*
BSSE: Basis set superposition error
CD: Circular dichroism
ChonABC: Chondroitin lyases ABC
CoM: Cofactor coenzyme M
COX: Cyclooxygenase
CS: Chondroitin sulfate
CYP: Cytochrome P450
DFT: Density functional theory
dNTP: 2'-deoxyribonucleoside 5'-triphosphates
DS: Dermatan sulfate
E. coli: *Escherichia coli*
EC: Enzyme commission
EchA: Epicholorohydrin epoxide hydrolase
EH: Epoxide hydrolase
EPR: Electron paramagnetic resonance
ER: Enoyl reductase
GIXD: Grazing incidence X-ray diffraction
GlcA: Glucuronic acid
HCN: Hydrogen cyanide
HDG: 1-O-hexadecyl glycerol
HF: Hartree-Fock
HIV: Human immunodeficiency virus
HNL: Hydroxynitrilase or hydroxynitrile lyase

HSV: *Herpes simplex* virus
IC$_{50}$: Half maximal inhibitory concentration
IR: Infrared
2-KPC: 2-(2-ketopropylthio) ethanesulfonate
2-KPCC: 2-ketopropyl-CoM oxidoreductase/carboxylase
K$_d$: Dissociation constant
LCP: Left circularly polarized
LOX: Lipoxygenases
MD: Megadalton
MM: Molecular mechanics
MP2: Møller-Plesset (perturbation theory)
NADPH: Nicotinamide adenine dinucleotide phosphate (reduced)
NMR: Nuclear magnetic resonance
NOS: Nitric oxide synthase
ONIOM: Our n-layered integrated molecular orbital and molecular
                 mechanics
OPA: One photon absorption
PFE: *Pseudomonas fluorescense* esterase
PM3: Parameterization 3 (of MNDO)
PKS: Polyketide synthase
PPi: Pyrophosphate
PTC: Peptidyl transferase center
PvulABCI: *Proteus vulgaris* chondroitinase ABCI
QM: Quantum mechanics
RCP: Right circularly polarized
R-HPC: [(R)-2-hydroxypropylthio]-ethanesulfonate
RNA: Ribonucleic acid
RT: Reverse transcriptase
SAG: Stearyl amine glycerol
TPA: Two-photon absorption
tRNA: Transfer RNA
TT: *Thermus thermophilus*
VCD: Vibrational circular dichroism
VOA: Vibrational optical activity
Yeast: *Saccharomyces cerevisiae*
ΔE_{LL-DL}: Chiral discrimination energy

chapter one

Introduction

A casual look is sufficient to draw our attention to the presence of symmetry in many of the objects around us. Apparent symmetry is observed in the animal kingdom, plants, as well as inanimate objects. Leonardo da Vinci's drawing of Vitruvian man reflects the symmetry present in the human figure. Bilaterally symmetric animals (other than humans) are also quite abundant and are examples of symmetrical natural objects. The reflection symmetry observed in the shapes of leaves and the radial symmetry present in the shapes of sea anemones are also common examples of symmetrical organisms. Symmetry is common among inanimate material objects. The impressiveness of architectural wonders such as the Taj Mahal, the pyramids, and the Parthenon is due to their unique symmetry. Obviously, the presence of symmetry creates an aesthetically pleasing sense related to the concept of proportionality and balance as well as the beauty of the object. The presence of symmetry in material objects not only appeals to the aesthetic sense but is also of practical use. Use of symmetry has been made for safety, security, and familiarity since ancient times. Use of symmetry rather than dissymmetry in the construction of a house or a room makes it convenient and livable for the inhabitants. Identifying an object in the latter case would need more effort than in the former case. This idea can be extended to various tools of day-to-day use. The reflection symmetry of bricks is used in construction, and helical symmetry is used in the working principle of drill-bits and springs. Numerous other examples can be given that indicate a preference for symmetry (both at the simple and complex levels) both in nature as well as in artificial habitats.

Is this apparent presence of symmetry ubiquitous or not? Notably, the concept of the presence of symmetry (or its absence) is related to the length scale of the object under consideration. A close inspection of the reflection symmetry in the shapes of leaves reveals that the leaf vein patterns show imperfect bilateral symmetry at the length scale of leave veins. Natural biological systems exhibit both symmetry and dissymmetry at various length scales of their structural details. The helical structure of the polypeptide shown in Figure 1.1 is a dissymmetric object. The dissymmetry of the polypeptide helix at the microscopic length scale is due to the presence of an asymmetric carbon atom (except glycine) in each of the constituent amino acids. On the other hand, the alpha helix structure of a

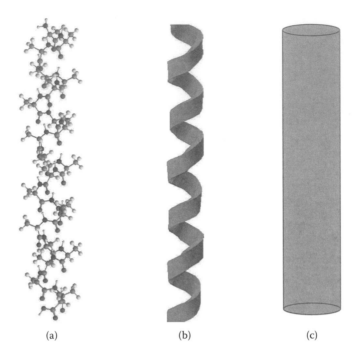

(a)   (b)   (c)

Figure 1.1 Length-scale-dependent symmetry and dissymmetry in a polypeptide chain using different levels of coarse-graining of the molecular structure. Different representations of a polypeptide are shown as (a) an all-atom representation exhibiting dissymmetry at the level of primary structure arising out of the chiral centers of individual amino acids, (b) helical representation showing the periodicity of the peptide linkage at the level of secondary structure, ignoring the atomistic structure, and (c) cylindrical representation showing the symmetry around the helix axis ignoring any underlying structural details. The dissymmetry is gradually lost in going from Figures 1.1a to 1.1c.

protein can be considered at the length scale of the complete helix as having cylindrical symmetry about its long helical axis (Figure 1.1). When the dissymmetry of the microscopic length-scale structure is ignored and the shape of the helical object as a whole is considered, the cylindrical symmetry of the object is apparent. In observing the symmetry of an object (or conversely, its absence), it is necessary to consider the relevant length scale. This length-scale dependence of symmetry is more apparent when one considers a functional biological molecule such as a protein. The possible point symmetry operations for protein quaternary structures are rotations since each subunit is a dissymmetric object, in general. The presence of the C_2 point group in α-chymotrypsin, oxyhemoglobin, and deoxyhemoglobin, the D_2 point group in concanavalin A, and the D_3 point group in insulin is recognized (Cantor and Schimmel, 1980, 127) despite the fact that all proteins at the microscopic length scale (length scale of amino acid or helical structure) are dissymmetric objects. Lack of symmetry at a certain level of the structural hierarchy is not only typical for proteins but is observed in many other biological molecules such as nucleic acids or lipids. Biological systems are overwhelmingly dissymmetric or chiral. Macroscopic, mesoscopic, and microscopic chirality can be noted in biological systems. Fundamental biomolecules such as amino acid and sugar are chiral. These molecules principally exist in one particular enantiomeric form and are examples of chirality at the primary or microscopic level. Since structure and function in biology are intimately related, a pertinent question is whether the chiralities of these natural objects have any influence on their functionality and if so, how they are related. We discuss this issue in the following text after a brief introduction to chirality and chiral discrimination.

1.1 Chirality and chiral discrimination

Chirality (from the Greek word *cheir*, meaning *hand*) is an attribute of nature that is observed in its various forms. Traditionally, chirality is considered a geometric property of various natural as well as nonnatural objects that makes them nonidentical and nonsuperposable with their own mirror images. A chiral object and its mirror image are enantiomorphic. On the other hand, an achiral object is superposable on its mirror image. This broad definition of chirality includes diverse macroscopic objects such as hands, animal organs, biological organisms, crystals, and macromolecules as well as microscopic objects such as various molecules as chiral entities (Eliel et al., 1994). If the object lacks an improper or alternating axis of symmetry, including a center of symmetry and a plane of symmetry, then it is chiral (Jacques et al., 1981, 3). A necessary and sufficient condition for the chirality of a molecule is that it cannot be superimposed on its mirror image as improper rotations are regular rotations followed by a reflection in the plane perpendicular to the axis of rotation. A molecule possessing an asymmetrically substituted carbon atom (four different groups as shown in Figure 1.2a) is a common example of chiral molecules and lacks all symmetry elements (other than the identity E or C_1). According to this definition, objects of various length scales, starting from the macroscopic objects mentioned earlier, to mesoscopic objects such as small- to medium-sized aggregates of molecules, as well as molecules of microscopic dimensions could be chiral entities. Chiral molecules such as amino acid and sugar are quite common natural objects. However, as exceptions, chiral molecules without asymmetric carbon atoms exist, and examples of achiral molecules with more than one asymmetric carbon atom are also well known. A detailed account of various aspects of molecular chirality is available in the standard literature (Eliel et al., 1994;

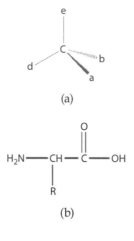

(a)

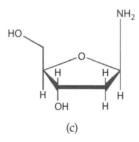

(c)

Figure 1.2 (a) General representation of a molecule possessing an asymmetrically substituted carbon atom to which four different groups—*a, b, d,* and *e*—are attached. This is a common example of a chiral molecule. (b) Amino acids are basic subunits of proteins and may contain a chiral center to which four different groups [a side chain (R), a carboxylic acid group (COOH), amino group (-NH₂), and a hydrogen (H)] are attached. Natural amino acids, except glycine, contain at least one chiral center. Amino acid is one of the most common examples of a chiral biological molecule. (c) Sugar is a basic subunit of nucleic acid and may contain more than one asymmetric center. Sugar is one of the most common examples of a chiral biological molecule that plays an important role in carrying genetic information.

Jacques et al., 1981; Flapan, 2000). Examples of chiral molecules are not only abundant among material systems such as organic and inorganic molecules but also among biotic systems. Amino acids and sugars contain asymmetric centers and are the most common examples of chiral biological basic subunits (Figures 1.2b and 1.2c, respectively).

    The broad term *chiral discrimination* encompasses differences in properties of different chiral objects such as those between enantiomeric and racemic objects. Chiral discrimination is a subtle phenomenon that arises, for example, when atoms or groups exchange positions in space (molecular

dissymmetry). The ensuing structural forms could be mirror image isomers, for example. The consequent differences in the three-dimensional
molecular structure of the enantiomeric and racemic objects can mani-
fest significantly different properties of the respective objects. From a
molecular viewpoint, chiral discrimination is a phenomenon where it is
possible to distinguish the intermolecular interaction of the molecules
of a homochiral nature (either R–R or S–S) with the interaction between
the molecule and its mirror image, which is of racemic type (R–S) or a
different chiral molecule. Examples are the differences in the structural
organization of enantiomeric and racemic systems at the molecular level.
However, the term *chiral discrimination* is not limited to the foregoing. The
difference between the absolute energies of L- and D- isomers (a small
energy difference) is considered chiral discrimination (Kuroda et al., 1978,
1; 1981, 33; Mason and Tranter, 1985, 45; Mason, 1989, 183; Mason, 1988, 347).
The differences between intermolecular energies of enantiomeric (L–L)
and racemic (D–L) pairs as a function of separation (a small but nonnegligible energy difference as a function of distance and a significant energy
difference as a function of mutual orientation) (Andelman, 1989, 6536;
Andelman and Orland, 1993, 12322; Nandi and Vollhardt, 2003c, 4033;
Nandi et al., 2004, 327) are also used as measures of chiral discrimination.
Since the broad definition of chirality includes objects of various length
scales, observables of macroscopic, mesoscopic, or microscopic length
scale are also used to study the chiral discrimination phenomena. The
differences in the volumetric properties of binary mixtures of optically
active liquid compounds (Leporl et al., 1983, 3520) and the major differences in the isotherm features of Langmuir monolayer composed of pure
enantiomeric (L–L or D–D) and racemic (D–L) monolayers are examples
of discriminations in macroscopic properties. The opposite-handednesses
of the monolayer domains composed of D- and L-enantiomers, presence
of both-handedness in racemic domains, opposite-handednesses of the
helical bilayer aggregates composed of D- and L-enantiomers, flatness of
the racemic bilayer aggregates in the condensed phase, and the differences in the lattice structures of the enantiomeric or racemic monolayers are examples of chiral discrimination at the mesoscopic length scale
(Andelman, 1989, 6536; Andelman and Orland, 1993, 12322; Nandi and
Vollhardt, 2003c, 4033; Nandi et al., 2004, 327).

Summarily, the manifestation of chiral discrimination could be
diverse, expressed as observables at macroscopic, mesoscopic, and microscopic length scales in biological, biomimetic, and material systems. A
schematic description of the chiral discriminations observed at various
length scales in a few common biological, biomimetic, and nonbiological
systems are shown in Figure 1.3. The length scales mentioned in the diagram should be considered to be approximate only since sharp boundaries do not exist for micro-, meso,- or macro-length scales.

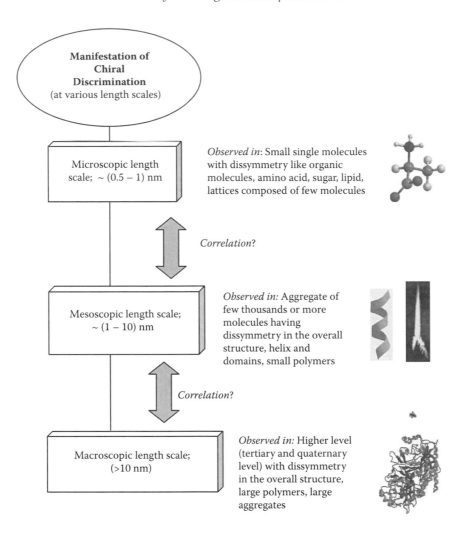

Figure 1.3 Schematic diagram showing the chiral discrimination observed at various length scales. The length scales mentioned in the diagram should be considered as approximate only as no rigorous boundaries exist for micro-, meso-, or macro-length scales (Nandi, N. 2009. *Int. Rev. Phys. Chem.*, 28: 111. With permission). For example, the domains formed in monolayers are often sized as several micrometers. The diagram principally refers to chiral discrimination in systems where electromagnetic interaction makes a dominant contribution to the discrimination. Hence, the discriminations arising from chiralities noted at the level of subatomic particles (less than 10^{-5} nm) and at the cosmic length scale are beyond the scope of this book and are not mentioned in the diagram.

Interestingly, the term *chiral discrimination* is not limited to the contexts mentioned above. For example, chiral discrimination appeals to the sense organs such as the smell generated by different enantiomeric molecules. The human olfactory system can discriminate between subtle differences in the chirality of the two enantiomers. The S-(+)-carvone has a caraway-like smell, whereas the R-(–)-carvone has a spearmint-like smell, and the latter is a stronger odorant on a threshold basis. Several other examples of different odors of enantiomeric molecules are known (Leitereg et al., 1971, 455; Nandi, 2003, 4588). The relationship between chiral discrimination and taste is also known (Shallenberger et al., 1969, 555). The present book deals with the phenomenon of chiral discrimination in which electromagnetic interaction plays the dominant role. Hence, the discriminations arising from chiralities noted at the level of subatomic particles (less than nanometer scale) and at the cosmic length scale are beyond the scope of this chapter (Hegstrom and Kondepudi, 1990, 108; Janoschek, 1991; Kondepudi and Durand, 2001, 351). The origin of chiral discrimination is, in general, difficult to comprehend because most properties (except those depending on chirality) and the formation energies (enthalpies and Gibbs free energies) are almost identical for the two enantiomers. Consequently, it is equally baffling to follow the molecular reasoning of the discrimination of the systems composed of enantiomeric and racemic molecules.

Due to the broadness of the scope of the term, chiral discrimination can be measured in several different ways. Enantiomers have many physical properties, such as melting point, boiling point, and density as identical. Properties such as optical activity depend on the interaction of the chiral molecule with an external electromagnetic field. Such properties are different for the enantiomers. The difference arises due to the differences in interaction of the electromagnetic field of the chiral molecule with the polarization state of the external electromagnetic field and results in discrimination in the coupling with either left- or right-circularly polarized light. The phenomenon has been known for a long time and is the subject of intensive study (Bose, 1898, 146; Mason, 1982; Craig et al., 1971, 165; Elezzabi and Sederberg, 2009, 6600). In principle, chiral discrimination can be studied experimentally by any method that is capable of revealing the differences in properties of enantiomeric and racemic systems, if any exist. Hence, a number of methods can be utilized, and we mention only those that have already been used to study discrimination in biomimetic or biological systems.

Optical properties that are dependent on the chirality of the molecules, such as optical rotation, electronic circular dichroism, and vibrational circular dichroism, have been used by chemists and biologists extensively to understand the structure and properties of chiral chemical systems (Berova et al., 2000; Harada and Nakanishi, 1972, 257; Taniguchi et al., 2009,

430). Circular dichroism (CD) is a conventional and very useful method of studying the electronic and stereochemical structure in chiral molecular systems as well as chirality-dependent vibrational dynamics. The differential absorption of left- and right-circularly polarized light by a material system is utilized in CD spectroscopy. In the absence of any externally applied fields, CD is observed only for systems that have at least some degree of handedness (or chirality) in their atomic or stereochemical structures. Early CD measurements have been confined to one-photon absorption (OPA) processes. However, CD measurements have later been reported for two-photon absorption (TPA) processes in a chiral system (Gunde et al., 1996, 195). Vibrational optical activity (VOA) of molecules is also a useful tool for studying chiral molecules. VOA is the differential response to left versus right circularly polarized infrared radiation (LCP versus RCP) during a vibrational transition. The IR form of VOA is known as vibrational circular dichroism (VCD). VCD can be used for many types of analysis related to the structure and conformations of molecules of biological interest, for example, in determining the enantiomeric purity of a sample relative to a known standard, in determination of absolute configurations, and the determination of the solution conformations of large and small biological molecules. The VCD spectrum of a chiral molecule is dependent on its conformation and absolute configuration. Thus, VCD allows determination of the structure of a chiral molecule. Chiral amphiphiles sometimes assemble into membrane structures with twisted, helical, or cylindrical tubular morphologies that express the chirality of their molecular constituents of micrometer or nanometer length scale, and VCD spectra are used for measuring discrimination (Berthier et al., 2002, 13486).

Although nuclear magnetic resonance (NMR) is a tremendously important and powerful technique in studying the properties of molecules in general, it is of limited use for the study of chirality. For example, the parameters that determine a spectrum, namely, the chemical shifts and spin–spin coupling constants, are identical for D and L enantiomers. Chiral discrimination has been achieved in NMR through the addition of a chiral reagent or solvent, producing a change in the environment of the active nuclei. Discriminations of various compounds are studied using this technique (Wenzel, 2007). Further improvements in estimating chiral discrimination using NMR are attempted. A nuclear magnetic moment in the x-direction combined with the strong magnetic field B^0_z induces an electric dipole moment in the y-direction through an odd-parity coupling in a freely tumbling chiral molecule. Estimates of the rotating electric polarization at the NMR frequency following a $\pi/2$ pulse indicate that it should be detectable in favorable cases (Buckingham, 2004, 1).

Several other experimental methods can reveal chiral discrimination, and the consideration of all possible methods is far beyond the scope of this chapter. For example, various experimental techniques such as surface pressure versus area per molecule (π–A) isotherm measurements, optical techniques such as Brewster angle microscopic (BAM) as well as fluorescence microscopic studies and grazing incidence x-ray diffraction (GIXD) studies are applied to biomimetic Langmuir monolayers. The properties of both enantiomers are measured and compared. The comparison has shown that the properties of chiral amphiphiles can be discriminated for enantiomeric and racemic systems. Conventional methods such as isotherm measurements, optical spectroscopy, and x-ray diffraction as applied to biomimetic systems are capable of revealing chiral discrimination in many cases at different length scales. Details of the methods applicable to chiral amphiphiles are available in the literature (Möhwald, 1990, 441; Knobler, 1990, 397; McConnell, 1991, 171; Vollhardt, 1996, 143; Nandi and Vollhardt, 2003c, 4033). Measurement of the surface pressure–area (π-A) isotherms at different temperatures was one of the early methods to probe chiral discrimination in Langmuir monolayers (Arnett et al., 1989, 131). Techniques based on reflection spectroscopy such as BAM and fluorescence microscopic techniques are also excellent methods to observe mesoscopic chiral discrimination effects in the monolayer. Using these techniques, one can draw inferences about molecular orientation by direct visualization of the monolayer. When p-polarized light is used and incidence takes place at the Brewster angle, no light is reflected from the pure air–water interface. However, changes in the molecular density or refractive index by the condensed phase of a monolayer on the aqueous subphase leads to a measurable change in the reflectivity and allows one to visualize and record the monolayer morphology. BAM studies can reveal many microscopic details of the monolayer without any external perturbation. BAM is advantageous compared to the fluorescence technique, which requires dye molecules as a probe to investigate the monolayer morphology. In the fluorescence microscopic studies, a fluorescent probe (a dye molecule) is added to the monolayer, which is excited using a high-pressure mercury lamp. Then the monolayer state is observed with a microscope mounted with a camera. A variety of probes have been used. BAM studies of monolayers revealed an enormous variety of domains in the condensed phase. Both x-ray and neutron beam techniques are employed to study the molecular organization of monolayers at the air–water interface. Specular reflectometry using x-rays or neutron beams probes the electron density or the neutron scattering length density profiles normal to the interface. The GIXD technique probes the molecular order in crystalline surface films. X-ray beams obtained from synchrotron radiation facilities are employed for this purpose. Discrimination at the length scale of lattice dimension is noted in many cases. Various biochemical methods are also capable of measuring the differences in the properties of enantiomeric and racemic systems from which the discrimination can be measured. Discussions on chiral discrimination in biomimetic and nonbiological systems appear in Section 1.1.1 and examples in biological systems are presented in Section 1.1.2.

1.1.1 Chiral discrimination in biomimetic and nonbiological systems

Extensive studies on chiral discrimination are available in the case of biomimetic (membrane-mimetic) systems such as monolayers and bilayers. Molecular chirality of the amphiphiles can induce as well as control the mesoscopic-length-scale chirality of the helix and domains in bilayers and monolayers, respectively. Differences in the interactions in the enantiomeric (L-L) and racemic (D-L) systems can give rise to the observation of significant discrimination (Nandi, 2004, 789). Bilayer aggregates (Nandi and Bagchi, 1996, 11208; 1997, 1343; Berthier et al., 2002, 13486; Dolain et al., 2005, 12943; Jiang et al., 2004, 1034) as well as monolayer aggregates composed of enantiomeric molecules (Nandi and Vollhardt, 2003b, 12; 2003c, 4033; 2006, 1; 2007, 351; 2008, 40) form three-dimensional (helix) and two-dimensional (curved domains) dissymmetric shapes with curvature, respectively. The curvatures or the handedness for L-L and D-D systems are opposite. Explicitly, in the case of the helical bilayer aggregate, if the L-L system develops a right-handed helix, then the corresponding D-D system will develop a left-handed helix. Similarly, in the case of monolayer domains, if the L-L system develops a domain curved in the left-handed direction, then the corresponding D-D domain will exhibit a domain curved in the right-handed sense. Racemic systems are flat (no handedness) in the case of the bilayer, and in the case of the monolayer domain may develop curvatures in both directions (reflecting mirror symmetry). These are examples of chiral discrimination in mesoscopic biomimetic systems. The effect of molecular chirality is manifested in monolayers and is clearly observable when the molecular separation is relatively small. Consequently, the chiral discrimination effects may not be observed in the gaseous or liquid-expanded phases of the monolayer. With increasing temperature or decreasing surface pressure, the rotational disorder of the molecules increases rapidly. The molecules rotate more freely about their long axis. As a result, the intermolecular interaction is insensitive to any asymmetry present about their short axis of the respective molecule, and the molecule becomes effectively symmetric to the neighboring molecules. Chiral discrimination effects are observed in chiral amphiphilic molecules containing various head groups such as amino acid, amide, glycerol derivatives, phospholipids, etc., based on π-A isotherm measurements Examples of chiral discrimination enantiomeric and racemic monolayers as noted in isotherm studies are shown in Figure 1.4.

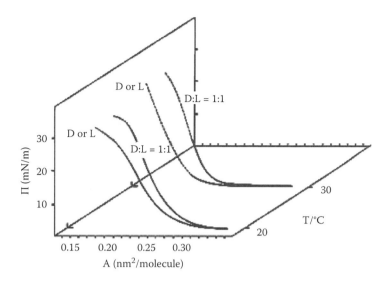

Figure 1.4 π-A isotherms of the enantiomeric and racemic monolayers of
N-dodecylmannonamide at 20°C and 30°C. Homochiral discrimination is indicated for both temperatures (Nandi, N. and Vollhardt, D. 2003c. *Chem. Rev.*, 103:
4033. With permission).

BAM and fluorescence studies revealed different domain shapes indicating discrimination. Examples of chiral discrimination noted in optical
studies, such as BAM and fluorescence studies, are shown in Figure 1.5
(Nandi and Vollhardt, 2007, 351; 2008, 40). Chirality has a major influence
on the morphology, which is clearly detected by optical techniques. As
was pointed out earlier, the effect of chiral discrimination is unobserved
in π-A isotherm measurements of enantiomeric and racemic systems in
several monolayers composed of chiral molecules, but clear discrimination is revealed from the direct visualization of the domain shape and
inner texture by BAM. Optical studies in the monolayers draw our attention to the fact that in many cases macroscopic methods such as the π-A
isotherm measurements fail to detect the chirality effects present at the
mesoscopic or microscopic level. A bewildering variety of inner structure
can be observed in domains composed of different classes of molecular
structure, also clearly revealed by BAM.

The molecular understanding of the discrimination in monolayers
has been developed in recent years (Nandi and Vollhardt, 2003b, 12; 2003c,
4033; Nandi et al., 2004, 327; Nandi and Vollhardt, 2006, 1; 2007, 351; 2008,
40; Nandi et al., 2008, 9489). When the monolayer of a chiral amphiphile is
composed of a pure enantiomer, the molecules within the corresponding
domains develop a mutual intermolecular azimuthal orientation. Whereas
the molecular tilt from the normal remains unchanged, the continuous or

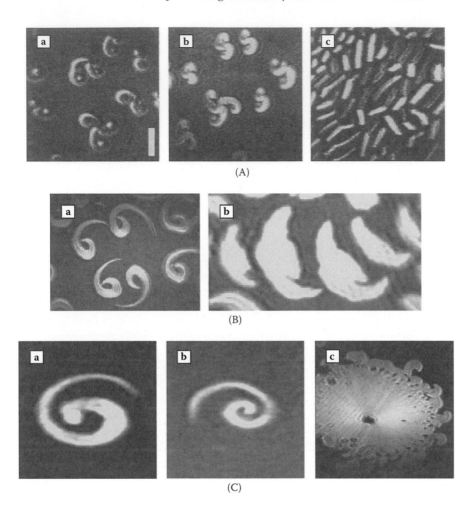

Figure 1.5 (A) BAM images of: (a) *D*-, (b) *L*-, and (c) *racemic*-dipalmitoyl phosphatidyl choline (DPPC) condensed phase domains. Bar length: 100 μm. (B) Chiral discrimination in the domain texture of *N*-α-palmitoyl-threonine monolayers: (a) D-enantiomer, (b) 1:1 DL-racemate. Image size: 350 × 350 μm. (C) Chiral discrimination of the condensed phase domains of stearoyl serine methyl ester (SSME) monolayers spread on pH 3 water: (a) D-enantiomer, (b) L-enantiomer, (c) 1:1 DL-racemate (image size 80 × 80 μm). (Nandi, N. and Vollhardt, D. 2007. *Acc. Chem. Res.*, 40: 351. With permission). (continued)

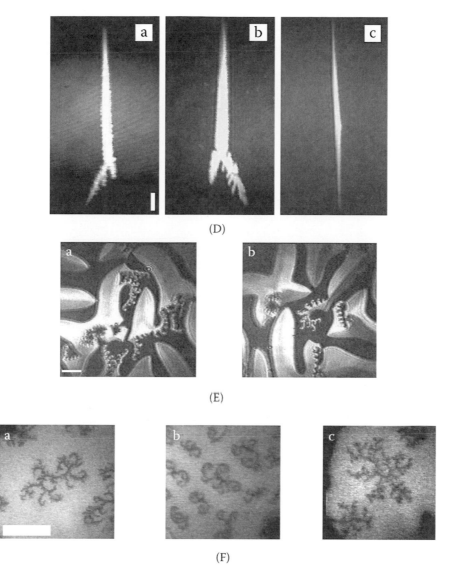

Figure 1.5 (continued) (D) Chiral discrimination of the condensed phase textures of tetradecyl dihydroxy pentanoic acid amide (TDHPAA) monolayer: (a) S-enantiomer, (b) R-enantiomer, (c) 1:1 RS racemate (bar length 50 μm). (E) Chiral discrimination in hexadecyl glycerol (HDG) domains: (a) racemic mixture with spirals curved in two opposite directions, (b) S-enantiomer with spirals curved only clockwise. (F) Chiral discrimination in 1-stearylamine glycerol monolayers: (a) S-enantiomer, (b) R-enantiomer, (c) 1:1 racemate (bar length: 100 μm) (Nandi, N. and Vollhardt, D. 2007. *Acc. Chem. Res.*, 40: 351. With permission).

sudden variation of the azimuthal orientation is a feature of the enantio-merically pure domains. The intermolecular orientation may vary along the length or width (or both) of the domain. This mutual orientation is propagated over the mesoscopic length scale. Consequently, the over-all domain shape or part of the domains exhibits curvature. In general, the racemic domains do not show anisotropic shape development. This indicates that either the mutual intermolecular azimuthal orientation is absent or canceled over the mesoscopic length scale in the case of hetero-chiral racemic systems. This is in contrast with the homochiral domains composed of enantiomeric molecules. However, chiral separation occurs in cases where the racemic domains develop curved arms with opposite handedness. This happens when the interaction between the same enan-tiomers (D–D or L–L) is favored over the racemic pair (D–L) (designated as homochiral preference). The opposite preference is termed heterochiral preference, where the interaction of the two different enantiomers (D–L) is more favored than the interaction between the same types of enantiomers (D–D or L–L).

However, the chiral discrimination in biomimetic systems can be very exotic, exhibiting both homo- and heterochiral discrimination in a single system. For example, the surface pressure of the main-phase transition of the pure enantiomers is higher than that of the racemic mixture, indicat-ing a heterochiral preference in 1-O-hexadecyl glycerol (HDG). On the other hand, the appearance of spirals with opposite curvatures at the same domain in racemic monolayers suggests chiral segregation, which means a homochiral preference. Similar observation of homo- and heterochirality is noted in stearyl amine glycerol (SAG) systems also. The simultaneous observation of surface-pressure-dependent hetero- as well as heterochiral-ity based on the domain morphology is rather different from the observed overwhelming homochiral preference in biosystems. The origin of homoch-irality and heterochirality in HDG and SAG is explained by computational studies (Nandi et al., 2004, 327; Nandi et al., 2008, 9489). The distance- and orientation-dependent crossover of homo- and heterochiral discrimination is also confirmed by experimental and theoretical studies of HDG mono-layers. Also, in that case, the theoretical calculation of the intermolecular pair potential for intermediate mutual separation indicates a heterochiral preference, whereas a homochiral preference is concluded to be gradually preferred for relatively shorter separations. The study explains the experi-mental finding of an interesting crossover of heterochiral preference and homochiral preference in a monolayer. On the other hand, the appearance of spirals with opposite curvatures within the same domains of racemic monolayers indicates chiral segregation of the two enantiomers, suggest-ing a homochiral preference above the transition pressure. Both homo- and heterochiral discriminations are not only dependent on the intermolecu-lar separation but also on their mutual orientation. Chirality-dependent

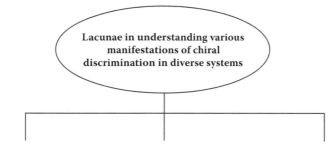

Lacunae in understanding various manifestations of chiral discrimination in diverse systems

Biological systems

- Why do the primary level biological subunits (amino acids, sugars, lipids) overwhelmingly exist in only one chiral form?
- What is the basis of the universal relation between the L-amino acid and D-sugar?
- What is the relation between the chirality at primary and higher levels (secondary and higher)?
- How is the enantiopurity retained in biological structures through evolution (mechanism and fidelity of the discrimination)?
- What are the influences of the primary level chirality on different vital biological reactions?
- How does the chiral discrimination occur in biological recognition events (odorant receptor-odorant, taste receptor-food; drug receptor-drug interaction)?

Biomimetic systems

- Why and how do monolayers composed of racemic mixture and enantiomeric molecules have different lattice structure, aggregate shape and isotherm properties?
- Why and how do bilayers composed of racemic mixture and enantiomeric molecules have different aggregate shape (flat and helical shapes, respectively)?

Nonbiological material systems

- What are the correlations between the monomeric units and the chirality of the higher level structure of polymeric structures (for example, helical chirality or propeller chirality)?
- Do these correlations bear the same principle as present in biological polymers?
- How can control be gained in "bottoms up nanofabrication" of specific structures with desired functions using chiral subunits?
- How can the achiral sub-units induce longer length scale chiral structures?

Figure 1.6 Schematic diagram showing a few questions pertaining to chiral discrimination observed in biological, biomimetic, and nonbiological material systems (Nandi, N. 2009. *Int. Rev. Phys. Chem.*, 28: 111. With permission).

interactions are not dominant at relatively larger separations (corresponding to the fluid phase), and in all cases, the calculated discrimination vanishes when the molecules are far separated.

The foregoing observations raise many questions (Figure 1.6). For example: (1) How does such a subtle molecular feature control longer length-scale structural features? (2) Why does a particular chemical structure (D or L form of a given amphiphile) give rise to its typical mesoscopic chirality (right- or left-handed domain)? (3) How in some cases does chiral segregation take place (despite the equal energies of D and L enantiomers)? and (4) How is intermolecular interaction to be treated in a situation where the customary assumption of isotropicity or symmetry over the length scale of the system (as commonly used in liquid-state theories) is not suitable? It is obvious that these questions cannot be explained by theories, which do not consider the molecular chiral structure and orientation-dependent intermolecular interaction.

In the condensed phase, the morphology of the bilayer aggregates composed of chiral amphiphilic molecules is strongly dependent on the chirality of the molecules. Bilayers formed by chiral amphiphiles can exist in states of various morphologies, such as helical ribbons, tubules, fibers, superhelical strands, flat crystals, or clothlike aggregates (Fuhrhop and Helfrich, 1993, 1565; Schnur, 1993, 1669; Selinger et al., 2001, 7157). The helical ribbons, tubules, and superhelical structures have a curvature that could be related to the chirality of the constituent molecules. If the amphiphilic monomer has at least one chiral center and the aggregate contains one kind of enantiomer (either D- or L-) in excess, it is observed that helical fibers and gels may be formed from the bilayers in the gel state. The helical structure formed from a particular kind of enantiomer has unique handedness. In the case of helical bilayers, the molecular chirality of the amphiphiles dictates the mesoscopic helical chirality of the aggregate. This observation is analogous to that in monolayers, where the handedness of the curved domain depends on the constituent amphiphiles. The corresponding racemic modification, on the other hand, does not produce helical fibers but only flat platelets or ribbons without a twist. Theoretical studies confirmed that the molecular chirality is the driving force for the formation of dissymmetric shapes (Nandi and Bagchi, 1996, 11208; 1997, 1343). Further discussions on the molecular understanding of chirality-dependent curvature are also available in the literature (Nandi and Vollhardt, 2006, 1).

Due to the diversity of chirality-induced morphology observed in biomimetic systems, computational studies have been used to understand the problem. It is possible to predict and understand the structure's formation from an effective intermolecular pair potential between the chiral centers of the monomers of the aggregate. Minimally, this potential should depend only on the distance and the orientation between the two participating chiral amphiphilic molecules (Nandi and Vollhardt, 2003c, 4033; 2007, 351; 2008, 40). The early studies made several approximations that are modified later. Theoretical work was carried out to explain the highly substance-specific orientational order of condensed phase domains observed in amphiphilic mono- and bilayers. The studies demonstrate that the intermolecular interaction profile is strongly orientation dependent for dissymmetric molecules and gradually loses symmetry with increasing dissymmetry of the molecules concerned (Nandi and Vollhardt, 2002a,

10144; 2002b, 207; 2000, 67; Nandi et al., 2002, 51; Nandi and Vollhardt, 2003a, 3464; Nandi et al., 2004, 327; 2004, 279; Thirumoorthy et al., 2006, 222). The dissymmetry in the pair potential profile indicates preferential orientations between molecules rather than an arbitrary mutual orientation. The preferred mutual orientation is the controlling factor of the curvature of the domains in the condensed state. Once the essential spatial dissymmetry of a given configuration is incorporated into the theoretical consideration, the observed manifestations of chirality at the condensed phase can be correlated with the microscopic structure by considering the low-energy conformations. Approximate models using various levels of detail of individual molecular structure (tetrahedral model, coarse-grained model, atomistic, or electronic structure based) are successful in describing the properties of the aggregate when the features of the corresponding chiral molecules are incorporated in calculations of the intermolecular energy profile. When the variation in the molecular conformations is large (for example, at an enhanced temperature), the effect of spatial dissymmetry can be drastically reduced because the intermolecular energy profile is less orientation dependent in that case.

The discrimination arises from the fact that when the positions of atoms or groups in space attached to the chiral centers of the D and L enantiomers are exchanged, it results in a change in the intermolecular interaction in the case of the enantiomeric and racemic pairs, respectively. The orientation and distance dependences of the intermolecular interaction profile of enantiomeric (L-L and D D) and racemic (D-L) pairs are different. The resulting difference in the orientations of the neighboring enantiomeric and racemic pairs drives the respective curvatures of the corresponding aggregate shapes. Thus, the subtle presence of molecular chirality controls the longer-length-scale architecture and induces different handedness of curvature. The observed orientation dependence acts over a mesoscopic length scale in the condensed phase, where the rotational degrees of freedom are restricted due to the dense packing of neighboring molecules relative to the fluid monolayer phase. The favored mutual orientation drives the neighboring enantiomeric molecules in the condensed state progressively arranged in a specific handedness and controls the features of the higher-level chiral structure (helices in the case of a bilayer and curved domains in the case of a monolayer). Such specific orientation dependence is absent in the corresponding racemic state, and no helicity or curvature is noted there.

Although the overall aggregate shape is guided by several factors such as interfacial tension (in three-dimensional aggregates such as bilayers) or line tension (in two-dimensional aggregates such as monolayers), and hydration, electrostatic interactions as well as chirality, handedness is essentially controlled by chirality. It is reasonable to draw the inference that the subtle stereogenicity at the chiral center of a chiral molecule is responsible for driving the aggregate shape to a particular morphology with a specific handedness. The molecular approach strongly indicates that the chirality-driven handedness of the bilayer helix or the sense of curvature of the domain in the monolayer is governed by the subtle stereochemical interactions at the chiral centers, which in turn are controlled by the pair potential between the groups attached to the chiral centers of the pair of amphiphiles. Theoretical studies on monolayers based on pair potential theories and similar studies on bilayers clearly established this fact (Nandi and Bagchi, 1996, 11208; 1997, 1343; Nandi and Vollhardt, 2006, 1). Chirality-dependent intermolecular interactions in other systems such as nucleic acid mimetic head groups and amino acids are also studied (Thirumoorthy and Nandi, 2005, 336; 2006, 8840).

It is now understood that the molecular chirality influences the intermolecular energy profile as a function of the distance and orientation between neighboring molecules in the condensed-phase aggregates of mono- and bilayers and other mimetic systems such as peptides. Computational studies show that the mesoscopic chiral shape of the condensed phase as well as other chirality-dependent features can be predicted from the molecular chiral structure by studying the intermolecular energy profile (Nandi and Bagchi, 1996, 11208; 1997, 1343; Nandi et al., 2004, 327; Nandi and Vollhardt, 2008, 40; 2006, 1; Nandi, 2004, 789). Discrimination in biomimetic model systems arises due to a nonnegligible difference in the intermolecular interaction profile as a function of mutual orientation and distance between L-L and D-L pairs. Theoretical studies also indicated that the discriminating nature of the profiles diminishes at large separation or higher temperature. Confinement is also an important factor for manifestation of chiral discrimination because chirality-dependent morphological features require that the molecules should be in close proximity, and the discrimination vanishes at larger separation as well as at higher temperatures. In an early theoretical study, Salem and coworkers concluded that the discrimination is small at freely rotating limits and only six-body or higher-order interactions will give rise to nonzero values for the relative difference between the two virial coefficients of L-L and D-L pairs (considered as a chiral discrimination parameter) at infinite temperature (Salem et al., 1987, 2887). This result is further supported by other theoretical studies (Andelman, 1989, 6536; Andelman and Orland, 1993, 12322).

In addition to the foregoing, various material systems exhibit micro- to macroscopic chiral discrimination. Chiral discrimination is noted in transition metal-catalyzed stereospecific polymers (Guerra et al., 2003, 1), diverse helical polymers (Rosa, 2003, 71), vinyl polymers, polyamides, polymers composed of polysaccharides and their derivatives (Okamoto et al., 2003, 157), polysilanes (Fujiki et al., 2003, 209), and various other polymeric materials (Green et al., 2001, 672). Similar to the biomimetic systems, the discrimination in the foregoing material systems is expected to be driven by the differences in the interaction profile of respective systems due to the stereogenicity of the chiral center of the concerned molecules.

The understanding gained in the case of biomimetic systems has implications for biological systems as chirality seems to be playing an important role there too. The complexity of biological systems requires systematic and casewise studies to understand the discrimination observed. We discuss the influence of chirality and chiral discrimination in biological systems in the following section.

1.1.2   Chirality and discrimination in biological systems

The structure–function relationship in biology implies that the influence of the chirality of biological building blocks at the primary level as well as chirality of the higher-level architectures must have an important function. For example, amino acids and sugars are two fundamental building blocks of the proteins and nucleic acids, respectively. These two molecules are connected with genetic information and the functional world of proteins, respectively. Various aspects of biological chirality are discussed in the literature in detail (Pályi et al., 1999, 3). The origin of the strong homochiral preference for L-amino acid and D-sugar at the primary level of biological architecture has been a long-standing problem (Pályi et al., 1999, 3). A related question is the origin of the heterochiral relationship between these two specific isomers of amino acids and sugars. This relationship was fixed sometime early in the evolutionary history (Berg et al., 2002, 41). Due to the similar energies of enantiomers, it is difficult to understand the preference of one enantiomer (either of amino acid or sugar) over the other.

Among the several unresolved questions related to the origin of life, the development of the homochirality of L-amino acid and D-sugar is definitely a challenging one (Figure 1.6). Some of the questions are mentioned in this figure. Paleontological studies seem most promising toward this end. It is indicated that the earliest enantiopure molecule (steroid derivatives) available are not older than 520 million years and functional molecules such as protein undergo racemization beyond 100 to 200 thousand years (Pályi et al., 1999, 3). Hence, the question of whether the biomolecules have been enantiopure for very long or since prebiotic time (for example, based on an extrapolation of present-day homochirality) is difficult to answer definitely. However, several suggestions have been made. A pre-RNA period or an RNA world is among the possible suggestions (Sandars, 2005, 49; Joyce and Orgel, 1999, 49). It has been pointed out that when self-replication in such a world was first established, the fidelity in the process was poor compared to the fidelity in today's world (Joyce and Orgel, 1999, 49). Further questions are yet to be answered concerning the features of chiral discrimination in biological systems. The genetic process, which includes coding and protein synthesis, is specific about the structural relationship between amino acids and sugars in nucleotides. A complementary relationship couples amino acid and nucleotides at both monomeric and polymeric levels. Heteropairing of amino acid (L-enantiomer) and sugar (D-enantiomer) in today's living world is best preferred by evolution. The preferred heteropairing of the L-amino acid and D-sugar over the other pairing possibilities indicates a biological chiral discrimination. The preference for other heteropairs (D-amino acid and L-sugar) and homopairs (either L-sugar: L-amino acid or D-sugar: D-amino acid) seems unlikely and is outcompeted by the natural heteropair. The specificity of the chiral heterocoupling of the enantiomeric forms of amino acids and sugars in functional biological systems indicates that the origin could be in the structure–function relationship of the heteropair. This specificity might have originated through prebiotic evolution and been retained thereafter. The origin and the mechanism of the selection pressure favoring the improvement of fidelity are also important questions. A prerequisite to any acceptable answer is that it must conform to the exclusive homochiral preference observed at the present time.

Different theories of the origin of homochiral evolution of biomolecules have been put forward. However, no conclusive answer is available to the question of how predominant homochirality occurred. Autocatalytic amplification of a very small chiral imbalance (Soai et al., 1995, 767; Avetisov and Goldanskii, 1996, 11435) has been proposed as a possible cause of homochiral evolution. However, it is difficult to refute some propositions or prefer others among different possible mechanisms of homochiral evolution because conclusions regarding the mechanisms are based on extrapolations back to prebiotic era. The studies of fossil records of chiral organisms (paleontological studies of chiral substances) might be useful in this regard (Fairbridges and Jablonski, 1979; Pályi et al., 1999, 3; Taylor and Taylor, 1993, 118; McGhee, 1989, 133). It is further interesting to note that exceptions to the predominant homochirality are also known. The presence of D-amino acids in several biological organisms is known, for example, in the nonribosomal peptide synthesis and posttranslational modification (Marahiel et al., 1997, 2651; von Döhren et al., 1997, 2675; Fuji, 2002, 103; Torres et al., 2007, 63; Stachelhaus and Walsh, 2000, 5775; Kreil, 1997, 337). A large number of biological molecules exhibit a strong preference for one of the enantiomeric forms out of the two (for each chiral center) nearly equal possible isomers. While incorporation of D-amino acids is a hallmark of peptide synthetase-based nonribosomal peptide synthesis (Marahiel et al., 1997, 2651; von Döhren et al., 1997, 2675–2705; Stachelhaus and Walsh, 2000, 5775; Challis and Naismith, 2004, 748) and posttranslational modification (Fuji, 2002, 103; Torres et al., 2007, 63; Kreil, 1997, 337), the RNA-dependent ribosomal synthesis of peptides and

proteins exclusively incorporates only the 20 natural amino acids and sele-
nocysteine with their natural chirality (L-form). Incorporation of D-amino
acids to an organism can be toxic if it does not naturally occur in the same
system (Friedman, 1999, 3457; Ohnishi et al., 1962, 138; Champney and
Jensen, 1970, 107; Cosloy and McFall, 1973, 685; Rytka, 1975, 562; Harris,
1981, 1031; Tsuruoka et al., 1984, 889; Caparros et al., 1991, 345; Caparos et
al., 1992, 5549). D-amino acid naturally occurs in selected systems such as
peptidoglycans of bacterial cell walls (D-Ala- and D-Glu-), peptide antibiotics, and in the human brain (D-Asp and D-Ser are present at high concentrations). Transformation of L-enantiomer by a racemase can also lead
to the development of D-enantiomer. Conversion of the L- to the D-stereo-
isomer of tryptophan was observed in the presence of tryptophan synthetase. The D-tyrosine might arise at the step of the addition of an amino
group to 4-hydroxyphenylpyruvate. Moreover, D-amino acids are likely
to be nonspecifically formed as side-reaction products in the presence
of pyridoxal phosphate-containing enzymes or of pyridoxal phosphate
alone. The presence and development of D-enantiomers only in some
specific organisms and almost complete exclusion of it in the natural biosynthetic pathway of protein synthesis in today's world raises the question of the mechanism of discrimination in different organisms. Since the
racemic compound in the crystalline state is more stable than the corre
sponding enantiomeric form (Jacques et al., 1981, 28), how is racemization
avoided throughout the course of evolution in the biosynthetic pathways?
The answer to the question of why the basic blocks of life are not being
scrambled and retain enantiomeric purity is challenging as many biological systems incorporate D-amino acids as well as racemize in other cases,
as mentioned before. Since nonnatural synthetic systems cannot match
the enantiopurity achieved by biosynthetic pathways, understanding of
chiral discrimination in biological systems is of obvious technological
importance.

At the secondary level of the structural hierarchy of these biological
molecules, helix and domains in proteins (Chothia et al., 1997, 597), helical deoxyribonucleic acid and ribonucleic acid secondary structures are
dissymmetric. Not only primary-level chirality, but also secondary-level
chirality is important for the functionality of the biological structure.
Chirality of the helix (which depends on the long-length-scale spatial
arrangement of residues) has a significant influence on the orientation-
dependent interaction with the chiral ligand (Nandi, 2004, 789). The interactions of both homopolymeric and heteropolymeric helices with achiral
ligands are orientationally isotropic but anisotropic with chiral ligand.
The interaction of a chiral ligand is most favorable only at certain orientations, which is not the case for an achiral ligand. Thus, chiral ligands are
expected to interact effectively with the helix in an orientation-specific
way compared to achiral ligands. The interaction with the corresponding

linear polypeptide is significantly less orientation dependent compared to the helical structure. It is also shown that the helix can recognize the enantiomeric ligand pairs based on distance- and orientation-dependent interactions. Significant discrimination in the interaction between a chiral ligand and its mirror image is noted. This supports the fact that enantiomeric odorants are recognized by G-protein-coupled transmembrane helices.

Naturally occurring α-helices composed of L-amino acids are more stable in a right-handed conformation. Left-handed helices are relatively rare (Novotny and Kleywegt, 2005, 231). This discrimination is due to the secondary-level chirality present at the length scale of the helix concerned. A recent detailed survey of left-handed helices (motifs of at least four consecutive residues that form at least one turn) indicates that they participate in the stability of proteins, constitute part of the active site, or play an important role in ligand binding. This is functionally interesting (Novotny and Kleywegt, 2005, 231). Further molecular-level studies are required to understand the relationship between the discrimination observed and properties as well as the function of the helical protein structure. This understanding could be useful in controlling the functionality and constructing the desired protein structure using a bottom-up approach. Tertiary- and quaternary-level structure of proteins (Dima and Thirumalai, 2004, 6564), supercoiled structures of nucleic acids, and large aggregates of lipids are examples of macroscopic dissymmetric structures. Little quantitative analysis is available concerning the role of such long-length-scale chirality and the function of related biomolecules.

It is important to note that the amplification to the total predominance of homochirality occurred and its retention in a racemizing environment continued after the symmetry-breaking event had taken place during the prebiotic time. What is the mechanism of retention that continues to operate in today's world? This question is not only of fundamental interest but is intimately related to the existence of life because the enantiomeric forms of sugar and amino acid constitute the very basis of life processes. Why the basic blocks of life such as L-amino acid and D-sugar are not being scrambled with their enantiomers but have retained enantiomeric purity in their biological architecture since evolution is an unresolved question.

A high level of stereospecificity in biological reactions has been noted for a long time. Since the retention of biological homochirality is vital for the functionality of biological macromolecules, it is important to follow the mechanism of chiral discrimination (mechanism of retention of enantiopurity in natural biosynthetic pathways) in microscopic detail. Since the molecules participating in vital biological reactions are often chiral, the influences of their molecular chirality obviously influence the discrimination in such reactions. These reactions occur within cavities or clefts that are parts of large biological macromolecules such as proteins, nucleic acids, or lipid aggregates. Common examples of such functionally important cavities or clefts are active sites present in enzymes and ribozymes as well as interfaces of lipid membranes. In the context of biomimetics, it was mentioned earlier that confinement and interaction with the neighboring chiral molecules are important factors for the manifestation of chiral discrimination. The nanodimensional active site cavities are organized structures composed of amino acid, nucleic acids (RNA and DNA), or lipids. It is expected that the chiral structure of the reactants enclosed in cavities and orientation-dependent interaction must have an important influence on the course of the reaction as well as chiral discrimination. The confinement of reactants within the nanodimensions of the active site is also responsible for the fidelity of the reactions. The reactants in the active site are enclosed in a restricted space during the reaction and enjoy far fewer translational and orientational degrees of freedom than are available in a bulk solvent. This confinement is expected to enhance the discrimination. It is known that the chemical structure and dynamics of the active site of enzyme greatly accelerate the corresponding biological reaction. The catalytic action in the active site makes the biological reactions much faster and surprisingly accurate compared to the same reaction carried out in vitro. In this book, we discuss how the chirality of the relevant molecular segments in the active site and the related intricate interaction influences the course of the reaction. Different chapters of this book deal with the recent understanding of the influence of chirality of amino acid and sugar in the active sites of oxidoreductases, transferase, hydrolases, lyases, ligases, and other enzymes.

1.2 Enzymes, active sites, and vital biological reactions

Living organisms could not survive without enzymes that catalyze vital life processes. A number of important chemical reactions take place every moment of the life cycle of a living organism. These reactions are facilitated by enzymes. Enzymes are functional molecular machines where stereoselectivity and the specificity of the chirality of the substrate are an accepted principle. Although it is notionally established that enzymes have the amazing capacity of chiral discrimination, a complete molecular understanding of such processes is yet to be achieved. Despite the fact that the remarkable stereoselectivities of enzymes have been known for a long time, a molecular understanding of this stringent chiral specificity was scanty. Biochemical, kinetic, and various other experimental studies on the stereospecificity of different enzymatic reactions are abundant in the literature and to provide a compendium of chiral discrimination exhibited by enzymes is beyond the scope of this book. On the other hand, this book attempts to present a perspective of the microscopic understanding of the influence of the active sites of different classes of enzymes on the process of chiral discrimination in the course of biological reactions carried out therein. It aims to collate understanding of the interactions present in the active site of enzymes that influence chiral discrimination. The interactions between the active site residues and the reacting species are discussed, and the concomitant influences on the discrimination, chiral specificity, and fidelity of the reaction are studied by experiment or computational techniques or both. Importantly, the study of the active site architecture and interactions therein indicates that the conserved residues might have specific roles in the reaction.

Enzymes can be classified based on the chemical reactions they catalyze. According to the Enzyme Commission number (EC number) scheme, various enzymes are classified in six classes (from EC1 to EC6). The EC scheme is a numerical classification for enzymes (Moss, 2006; Webb, 1992). Oxidoreductases (alternatively called *dehydrogenase* or *reductase;* the name *oxidase* is only used when oxygen is the acceptor) belong to class EC1 and catalyze oxidation-reduction reactions involving transfer of protons, oxygen atoms, or electrons from one substance to another. The EC classification for these enzymes classifies them firstly by the *donor* group that undergoes oxidation, then by the *acceptor*, and lastly by the *enzyme*. Transferases (class EC2) carry out the transfer of a functional group (for example, methyl-, acyl-, amino, glycosyl, or phosphate group) from one substance (which is generally regarded as the *donor*) to another. The recommended names are normally formed as follows: *acceptor-group transferase*, or *donor group transferase*. The EC classification for these enzymes classifies them first by the *donor* group, then by the *acceptor*, and lastly by the *enzyme*. Hydrolases are classes of enzymes (class EC3) that form two products from a substrate by hydrolysis reaction. The hydrolases catalyze the hydrolytic cleavage of C-O, C-N, C-C, and some other bonds, including phosphoric anhydride bonds. The EC classification for these enzymes generally classifies them first by the nature of the *bond* hydrolyzed, then by the nature of the *substrate*, and lastly by the *enzyme*. Lyases (class EC4) perform nonhydrolytic addition or removal of groups from substrates involving change in C-C, C-N, C-O, or C-S bonds). The lyases are enzymes cleaving C-C, C-O, C-N, and other bonds by elimination, leaving double bonds or rings, or conversely adding groups to double bonds. Isomerases (class EC5) are enzymes that make intramolecular geometric or structural rearrangements leading to isomerization of a given molecule. According to the type of isomerism, they may be called *racemases, epimerases, cis-tran-isomerases, isomerases, tautomerases, mutases,* or *cylcoisomerases*. The subclasses are defined according to the type of *isomerism*, the sub-subclasses of the type of *substrates*. Ligases (class EC6) synthesize new C-O, C-S, C-N, or C-C bonds with simultaneous breakdown of ATP to join two molecules together. The ligases catalyze the joining together of two molecules coupled with the hydrolysis of a pyrophosphate bond in ATP or a similar triphosphate. The bonds formed are often high-energy bonds. The subclasses are defined according to the type of *bond* formed; the sub-subclasses are only used in the C-N ligases.

Enzymes carry out the reactions within the respective active sites. Active sites of enzymes are the reaction vessels of nature, where molecules are synthesized as well as being broken to cater to the various needs of life. From a structural and chemical perspective, the active site generally refers to a cavity or a cleft within a functional biological molecule such as an enzyme catalyzing a biological reaction. It is elegantly pointed out by Warshel and coworkers that it is insufficient to just state that an enzyme binds the transition state stronger than ground state and the catalytic groups are properly oriented to enhance the reaction rate (Warshel, et al., 2006, 3210). It is important to know how the more effective binding of the transition state is achieved and the catalytic groups involved in the process. The major issue is to understand why the enzymatic reaction is much faster within the enzyme active site than the corresponding reaction in solution. It is certain that molecular chiralities of substrates and enzymes play a significant role in the process.

1.2.1 Active sites as reaction centers

The closest vicinity of the reactants within an enzyme is the respective active site. The different regions of the active site and the respective residues enclose the reactants, locate them in the proximal position as well as orientation (suitable for reaction), and catalyze the reaction. A distinction between the reaction chemistry and the catalytic mechanism has already been pointed out in the literature (Todd et al., 2001, 1113). *Chemistry* refers to the overall strategy of changing substrate into product, and the nature of the intermediates involved, whereas *catalytic mechanism* describes the roles played by specific residues in the active site. It is rightfully pointed out that the "catalytic effect is entirely due to the active site environment" (Warshel et al., 2006, 3210). Consideration of the chirality of the active site residues is necessary to address this issue since the different spatial arrangement of their groups can make diverse orientation-dependent interactions, which are effective for binding of different substrates. The activation free energies in enzyme active sites calculated by computational methods such as empirical valence bond and quantum mechanical/molecular mechanical methods indicate that the catalytic power of enzymes is exclusively due to electrostatic effects. The preorganized electrostatic environment of the active site is the origin of the catalytic power of the enzyme (Warshel et al., 2006, 3210).

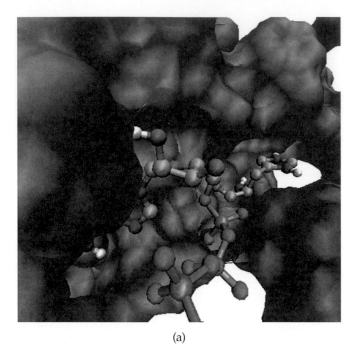

(a)

Figure 1.7 **See color insert.** (a) Schematic representation of the active site of prokaryotic *Escherichia coli* Histidyl tRNA synthetase generated from the crystal structures 1KMM.PDB and 1KMN.PDB (Arnez et al., 1997, 7144). The substrates (histidinol and ATP) are shown by a ball and stick representation and the enclosing nanospace (cavity of the active site) is shown as a gray surface. The images are prepared using VMD (Humphrey, W. et al., 1996. *J. Mol. Graphics*, 14: 33). (continued)

The walls of the active site cavity are composed of catalytic and other residues, often made up of amino acids or nucleotides. The active site of proteinous enzymes is composed of amino acids. Schematic representations of the active site of prokaryotic *Escherichia coli* Histidyl tRNA synthetase, composed of amino acids, are shown in Figures 1.7a and 1.7b, respectively. Active sites can also be composed of nucleic acids. The peptidyl transferase center (PTC) is the active site where peptide bond formation occurs. This site is located in the large subunit of the ribosome and is composed of layers of conserved nucleotides (Youngman et al., 2004, 589). Schematic representations of the active site of the peptidyl transferase center taken from the crystal structure of the ribosomal part of *Haloarcula marismortui* are shown in Figures 1.8a and 1.8b, respectively.

The active site structure and the chemical properties of the residues therein are responsible for the recognition and substrate specificity. The residues bind the reactants to a specific location at a suitable orientation

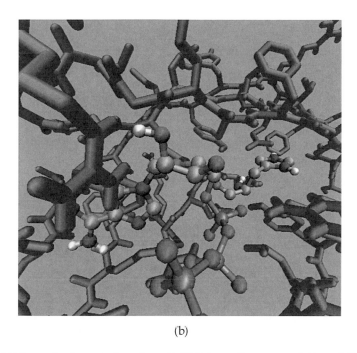

(b)

Figure 1.7 **See color insert.** (continued) (b) Schematic representation of the active site of prokaryotic *Escherichia coli* (same as in Figure 1.7a) in which the amino acids constituting the enclosing nanospace (cavity of the active site) are explicitly shown with the stick representation. The images were prepared using VMD (Humphrey, W. et al., 1996. *J. Mol. Graphics*, 14: 33).

to carry out the reaction efficiently. Substrates bind to the active site of the enzyme mostly through nonbonded interactions: electrostatic, hydrogen bond, hydrophobic interactions as well as van der Waals interactions to form the enzyme–substrate complex. The vital importance of the active site is that it modifies the reaction mechanism by decreasing the activation energy of the reaction compared to that of the same reaction in bulk. According to the lock-and-key model, the active site is a perfect fit for a specific substrate and once the substrate binds to the enzyme no further modification is necessary. On the other hand, the induced fit model assumes that an active site is flexible, and conformational changes may occur as the substrate is bound. The flexibility in the active site should not be interpreted as a strain-free state. Indeed, the active site is a preorganized state and is highly constrained in order to stabilize the transition state (Warshel, 1978, 5250; Warshel et al., 2006, 3210; Marti et al., 2008, 2634). Once the reaction is complete, the reaction product is subsequently released from the active site, and the enzyme returns to its initial unbound state.

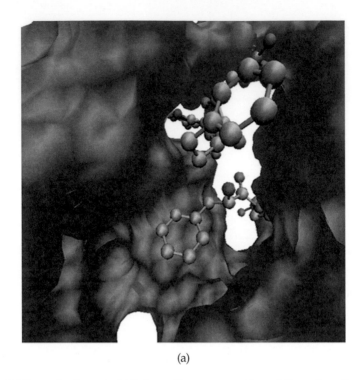

(a)

Figure 1.8 **See color insert.** (a) Schematic representation of the active site of architecture of the peptidyl transferase center (PTC) taken from the crystal structure of the ribosomal part of *Haloarcula marismortui* (Hansen et al., 2002, 11670). The crystal structure of CCA-Phe-cap-biotin bound simultaneously at half-occupancy to both the A-site and P-site of the 50S ribosomal subunit (1Q86.PDB) is used to generate the structure. The reactants (amino (A) and peptidyl (P) terminals) are shown by a ball and stick representation and the enclosing nanospace (cavity of the active site) is shown as a gray surface. The images are prepared using VMD (Humphrey, W. et al., 1996. *J. Mol. Graphics*, 14: 33). (continued)

The active site is only a part of a large macromolecule such as a protein composed of several residues or an RNA molecule composed of numerous nucleotides. While the overall enzyme structure participates in the catalytic process (due to the long-range electrostatic influence of distant residues on the reactants located within the active site), active site residues are closest to the substrate and hence have a dominant influence on the course of the reaction. The structure of the active site is important in terms of the presence of catalytic residues in the immediate vicinity of the reactants, positioning of the reactants suitable for reaction, as well as the reduction in the electrostatic potential to effectively carry out the reaction. This book primarily focuses on the influences of chirality of amino acids or sugars on various enzymatic reactions in the active site where

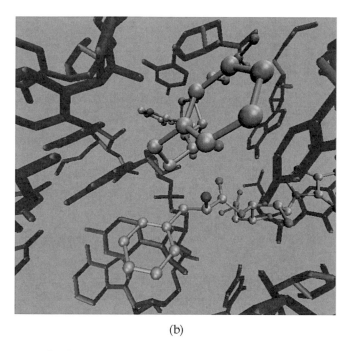

(b)

Figure 1.8 **See color insert.** (continued) (b) Schematic representation of the active site of architecture of the peptidyl transferase center (PTC) (same as in Figure 1.8a) in which the nucleotides constituting the enclosing nanospace (cavity of the active site) are explicitly shown with the stick representation. The images were prepared using VMD (Humphrey, W. et al., 1996. *J. Mol. Graphics*, 14: 33).

chiral discrimination is noted. It is obvious that the chirality of the nearest neighbors has a dominant influence on chiral discrimination in enzymatic reactions due to $(1/r^n)$ dependence of interactions of electromagnetic origin. Since the active site residues are close to the reactants, it is instructive to look into their organizational pattern, that is, how they impart the influence of their chirality to the reactants. However, such a consideration does not imply that the overall structure of the enzyme is irrelevant in understanding the influence of chirality on the enzymatic reaction. It also does not intend to provide an incomplete view of the chiral discrimination in such reactions. Since a protein's local structure and dynamics are coupled with the longer-length-scale structure and dynamics, the active site structure and dynamics are correlated with the structure and motion of the higher-level (secondary or tertiary level) structures. When a distant residue (distant from the active site) is mutated, it can change the active site structure as well as modify the dynamics (being composed of residues that are parts of secondary or tertiary structures and are projected on the active site walls). These changes can influence the interaction between the active site structure and the reactants therein. Whether a particular mutation can have such an effect or not depends on the specific mutation made and the particular secondary and tertiary protein structure. Only a molecular-level analysis of the changes in the interaction and dynamics studied before and after mutation can reveal the observed influence on the ensuing enantiodifference.

It was discussed recently that a rational approach toward improved enzymatic activity can significantly enhance enantioselectivity that focuses on mutations close to the active site (Horsman et al., 2003, 1933). On the other hand, directed evolution (recursive mutagenesis and screening for improved enzyme or enzyme mimetics) suggests that mutations far from the active site control enantioselectivity. However, the precise reason for such enhanced enantioselectivity is unclear. Horsman and coworkers observed that directed evolution can moderately increase the enantioselectivity of *Pseudomonas fluorescence* esterase (PFE). Although it is observed that distant mutations increase enantioselectivity, a possible molecular basis for the increase in enantioselectivity is due to indirect changes to the substrate-binding site. In the absence of available crystal structure, homology model building is used to propose that the mutation of distant residues might influence the substrate-binding site. It is proposed that mutations targeted to the active site would be a more effective strategy for increasing enantioselectivity. Only about one third of random mutations involve an amino acid whose Cα lies within 15 Å of the stereocenter carbon. Due to the greater number of residues present far from the active site, random mutagenesis is strongly biased toward mutations far from the active site rather than closer to it. It has been pointed out that amino acid residues within 10 Å of the active site comprise about 10% of the total amino acid residues present in a protein. However, Horsman et al. observed that all three mutations identified to increase enantioselectivity in this work lie closer to the active site than expected by random mutation. These results suggest that mutations close to the active site would be even more effective in increasing enantioselectivity than random mutations.

It is further pointed out that the closer mutations are more effective than distant ones in many cases (but not in all cases) (Morley and Kazlauskas, 2005, 231). Mutations closer to the active site can lead to improved enzymes in terms of substrate selectivity and catalytic activity. This is consistent with the concept that the active site residues have a significant role in chiral discrimination. Instead of mutating the entire enzyme, focusing mutations near the substrate-binding site might dramatically increase the success rate in many directed evolution experiments. A large number of studies are devoted to design of enzymes (Benson et al., 2000, 6292; Pérez-Payá et al., 1996, 4120; Bryson et al., 1995, 935; Kamtekar et al., 1993, 1680; Benner, 1993, 1402; Handel et al., 1993, 879; Xiang et al., 1999, 7638).

It has been suggested that a first-principle description of the reaction in terms of entire reaction coordinates as well as considering the dynamic protein structure by recognizing it as a flexible molecule would lead to efficient catalytic design. It is also pointed out that mutation of nonactive site residues might influence enantioselectivity. Other properties of enzymes such as thermostability or solvent tolerance could be dependent on nonactive site residues (Otten et al., 2010, 46).

It is shown that residues distant from the active site can significantly influence substrate specificity by their cumulative effects. The three-dimensional structure of the mutant enzyme indicates that the active site is remodeled, the subunit interface is altered, and the enzyme domain that encloses the substrate is shifted by the mutations (Oue et al., 1999, 2344). Clearly, the roles of the distant residues are modifying the active site structure or changing the interaction with the active site or catalytic residues with the substrate. Due to their remote position, they are unable to directly participate in the reaction mechanism. Hence, the active site residues have a more significant role in chiral discrimination. As they are less targeted by mutation by their sheer lesser abundance compared to distant residues, their influence is less noted.

1.2.2 Active site structures in enzymes

It is not easy to locate an active site that is the place containing residues for the reaction from among numerous other residues within an enzyme which may contain several cavities or clefts within a macromolecule such as ribosome. Efforts have been made to identify the active site and residues among several other residues in enzymes (Taylor, 1986, 205; Lichtarge et al., 1996, 342; Laskowski et al., 1996, 2438; Schneider, 1997, 427; Ondrechen et al., 2001, 12473; Armon et al., 2001, 447; Elcock, 2001, 885; Landgraf et al., 2001, 1487; Madabushi et al., 2002, 139; Kortemme and Baker, 2002, 14116; Pei et al., 2003, 11361; Gutteridge et al., 2003, 719; Jones and Thornton, 2004, 3; Chelliah et al., 2004, 1487; Wang and Samudrala, 2005, 2969; Cheng et al., 2005, 5861). Enzymes are chosen primarily by the respective EC number for analysis of the active site comparison. Such data are obtained from the Enzyme Structures Database (http://www.biochem.ucl.ac.uk/bsm/enzymes/index.html and http://www.ebi.ac.uk/thronton-srv/databases/enzymes/) and are retained in the data set if there is an available x-ray crystal structure or NMR model. Further considerations are given to the data about the active site, overall reaction catalyzed, and catalytic mechanism if any such information is available from the primary literature. Active site residues are responsible for substrate specificity and recognition or binding as well as catalysis. A feature of these catalytic residues is that they are highly conserved in sequence. It is plausible that the conservation is a consequence of the evolution targeted to develop and retain efficiency in the specificity and fidelity of the reaction. There are many complications in assigning the function of a catalytic residue, due to the multistep nature of chemical reactions. One residue can play more than one role and can be involved in different steps of the reaction. In principle, catalytic mechanisms can only be quantitatively modeled by quantum mechanical methods, by assigning the changes in the electronic structure involved during the course of the reaction.

Data sets of nonhomologous enzymes of known structures with a well-defined active site and plausible catalytic mechanism are available in the literature (Bartlett et al., 2002, 105). Classification of catalytic residues in various enzymes is available in the literature, which considers those residues that are thought to be directly involved in some aspect of the reaction carried out by the specific enzyme. Catalytic residues of the active sites of enzymes of known structure and catalytic mechanism are defined by manual inspection of the primary literature. A catalytic residue might be directly involved in the catalytic mechanism (may act as a nucleophile), can alter the pK_a of a residue or water molecule directly involved in the catalytic mechanism, can stabilize a transition state or intermediate, and lowers the activation energy for the corresponding enzymatic reaction or might activate the substrate (by polarizing a bond to be broken). Residues involved in ligand binding are excluded unless they also fulfill one of the foregoing criteria (Porter et al., 2004, D129). The functions of the active site catalytic residues are also specified in the literature, which includes acid-base catalysis, nucleophilic catalysis, transition state stabilization, activating water molecules directly involved in reaction, exerting a favorable influence and activating cofactor, activating another residue involved in the catalytic mechanism or a cofactor or substrate, forming a radical involved in catalysis of a reaction, or being modified in some way so as to perform catalysis during the reaction (Bartlett et al., 2002, 105).

Proteinous enzymes belong to diverse superfamilies, which are also classified as a group of evolutionarily related proteins, descended from a common ancestor typically having similar three-dimensional structures, functions, and significant sequence similarity. Enzyme superfamilies exhibit remarkable variations concerning conservation and variation of substrate specificity, reaction chemistry, and catalytic residues. Detailed analysis of substrate specificity has been carried out (Todd et al., 2001, 1113). They display remarkable variations in substrate selectivity. It is noted that out of 28 superfamilies involved in substrate binding, the substrate is absolutely conserved in only one, and enzymes in six super-families bind to a common substrate type, such as DNA, sugars, or phosphorylated proteins. At least in three of these superfamilies, variations within these ligand types may be extensive. Accordingly, their active sites differ vastly in shape and size. Substrate diversity implies diverse binding sites, achieved through structural variations and exploiting the different structures of the 20 amino acids present in the active site. Chiralities of the different amino acids are manifestations of the structural diversities of the respective amino acids and have known influence on the reaction chemistry involving these amino acids in bulk. It is expected that chirality should have a significant influence on reactions occurring within the active site.

The correlation between the conservation of active site residues in different species and the molecular mechanism of a particular reaction is yet to be understood in detail. The enzyme superfamilies exhibit remarkable variations in substrate specificity, reaction chemistry, as well as catalytic residues (Todd et al., 2001, 1113). Some enzymes are similar in their substrate specificity as well as the active site residues (catalytic residues) and follow the same reaction chemistry. Few classes of enzymes display functional diversity despite the presence of the conserved catalytic residues (employ a similar chemical strategy for catalysis). On the other hand, a number of superfamilies are noted in which the catalytic residues are nonconserved. It is possible that the catalytic residues may be located at different points within the protein fold or might involve different residues in catalysis. A detailed study concerning these aspects of active site organization is available in the literature (Todd et al., 2001, 1113). Since the mechanism of a given enzymatic reaction is crucially dependent on catalytic residues, which are known to lower the transition state barrier height as well as control the fidelity of the reaction, the related question is how the structures of the active sites (where the reaction takes place) compare in different species. It is also an open question how the conservation as well as diversity of residues in the active site space influences the molecular mechanism of a particular enzymatic reaction. This is equivalent to asking the question: do active sites with similarity in function have some common structural pattern and does the chirality of the architecture (at both microscopic and higher length scale) of active sites in these enzymes have any influence common in all superfamilies? What is the relationship between the structural pattern (if it exists) and the reaction (and related function) carried out therein? A careful look at the enzyme families is necessary from the viewpoint of chirality of the constituent molecular segments. According to a study, out of more than 30 superfamilies studied, a similarity in reaction chemistry is noted in a majority. However, it is indicated that conservation is noted in 4 superfamilies, and 22 are semiconserved (Todd et al., 2001, 1113). Conservation of principal catalytic residues is noted in 14 superfamilies, in 11 cases the residues are not conserved, and others are either semiconserved or unknown. It is observed that a given active site can catalyze a host of diverse activities, and conversely, different catalytic frameworks in related proteins can execute very similar functions (Todd et al., 2001, 1113). Thus, both the conservation and divergence of the function and active site architectures are observed.

It is useful to ask how the active site structure is conserved. Three-dimensional patterns of the catalytic residues can be strikingly conserved between distantly related enzymes. They can also be very similar in unrelated enzymes of similar function. A typical example is the Ser-His-Asp catalytic triads found in a variety of hydrolases (Torrance et al., 2005, 565). An understanding of the extent of this structural conservation in catalytic sites can be utilized for enzyme function prediction. Searching for these residue patterns is a useful complement to methods based on sequence or overall structure, for several reasons. There are instances when proteins have independently evolved the same configuration of catalytic residues for carrying out similar reactions. In these cases, it may be possible to predict the common function on the basis of the common catalytic residue conformation. It is also possible that the catalytic residue conformation in homologous enzymes of similar function may remain conserved but the rest of the protein structure has diverged, leading to difficulty in predicting function. Even when distant homologues can be identified using sequence methods, their correct sequence alignment may be ambiguous. Identification of similar catalytic sites that are distributed over various protein chains may be easier using structural similarity of catalytic sites than by using sequence comparison.

It is also important to understand how much structural variation exists between the catalytic sites of enzymes of similar function and the reasons for the variation that is present. The answer may be related to the molecular mechanism of the reaction and active site structure needed to efficiently catalyze the process in a better way. The template library was used to analyze the extent of variation of catalytic sites within enzyme families (Torrance et al., 2005, 565). One probable origin of the difference between catalytic sites is evolutionary divergence. The variability may reflect evolutionary divergence and optimization of the catalytic efficiency of these enzymes. Similar reactions that involve different substrates are likely to vary in the precise details of their reaction chemistry, such as in their way of transition-state stabilization. It is suggested that the new functional groups accidentally evolved in the active site might have taken over the roles of earlier residues if the newer residues perform the reaction more efficiently.

Several approaches are directed toward the study of biological reactions in active sites. An important approach is the modification of the structures of the active site residues. Two conceptually similar but experimentally different methods have been used for the specific modification of amino acid functional groups present in active sites. One classical method is to use chemically reactive small molecules that are designed to react with a restricted class of functional groups. The discrimination for a specific type of group in the active site is determined by the specific chemical reactivities of functional groups present in the active site. A large number of reagents are available for the modification of thiolate, imidazolate, hydroxylic, phenolic, and carboxylate functional groups present in active sites. The enhanced affinity of the functional groups to either electrophilic or nucleophilic reagents is an important factor for the method to be successful. However, the lack of absolute specificity for active site functional groups, the uncertain impact of the steric and electrostatic properties of the modified amino acid on the mechanism of the reaction (in terms of both the catalysis and binding specificity), and the effect of the modification on the three-dimensional structure of the covalently modified variant enzyme could be the limitations of the method. The approach of chemical modification of functional groups in the active site is very specific. The method need not involve the introduction of sterically or electrostatically important additions to the active site. The electrostatic effect being long-ranged, results obtained from such modifications must be judged with caution. On the other hand, the method is limited by both the possible number of substitutions for an amino acid and by the difficulty in unambiguous assessment of the effect of the amino acid substitution on the structure of the modified enzyme. The other approach is the genetic change of the amino acid functional groups present in the active site of an enzyme by specific alteration of the codons for these amino acids in a cloned or chemically synthesized gene for the enzyme. This method is based on recombinant DNA techniques and referred to as site-directed or site-specific mutagenesis (Gerlt, 1987, 1079).

Knowledge of the structure and improved understanding of the properties of enzyme active sites as well as their catalytic mechanisms are vital for novel protein design and predicting protein function from structure. Crystallographic and NMR studies of enzymes have shed light on the relationship between an enzyme's three-dimensional structure and the chemical reaction it performs. However, from the structure alone, it is a challenging task to extrapolate a catalytic mechanism. Detailed biochemical information about the enzyme can be used to design substrate or transition state analogues, which can then be bound into the enzyme for structure determination. These can reveal binding site locations and identify residues, which are likely to take part in the chemical reaction. From this, a catalytic mechanism can be proposed and can be confirmed by other information, for example, site-directed mutagenesis, kinetic analyses, and by extrapolation from homologues.

1.3 Chirality and reactions in active sites

1.3.1 Chiral discrimination and biological reactions

Studies on chiral discrimination in biomimetic systems indicated that the specific orientation and positioning of the chiral amphiphilic molecules is important for the manifestation of discrimination. It is mentioned that the discrimination in biological molecules is stringent. It might be possible that the reactions which lead to the synthesis of such molecules or involve them might have specific orientation dependence, relative position, and the pattern of the chiral moieties near the active site. This suggests that the chirality of active site residues might be correlated with the chiral specificity of reactions (Nandi, 2009, 111). Studies in biomimetic systems also indicate that chiral discrimination can be significant when the molecules are confined (by other molecules, for example) within nanometer range. The reasons for such augmentation are studied in detail for biomimetic monolayers and are mentioned before. Since the reacting moieties are confined within the active site, it is possible that confinement within a nanosized cavity might have an important influence on discrimination (Nandi, 2009, 111). Chiral discrimination vanishes if the chiral molecules are separated over a longer scale (tens of nanometers or larger). The measure of separation would be dependent on the nature of the groups attached to the chiral center. If the groups are nonpolar (such as alkyl groups), then the interaction involved is of van der Waals type, which operates over a smaller separation. Ionic interaction or dipolar interaction acts over tens of nanometers, as in the case of charged or partially charged groups. Consequently, the orientation dependence of interaction and the effect of chirality and discrimination depend on the nature of the group attached. However, chiral discrimination is expected to be insignificant when the separation is rather large (several tens of nanometers or larger separation). It is thus possible that the chirality of the active site residues (close to the reacting moieties) might have a more important role in the discrimination in reaction than the residues far from the active site.

Active sites composed of chiral moieties such as amino acids and sugars can have diverse specificity depending on the different mutual spatial arrangements of these residues. This indicates that diverse stereospecific reactions can be effectively carried out within such sites composed of chiral biological subunits such as amino acids and sugars. Rearrangement of a small number of chiral moieties is sufficient to obtain a diverse interaction profile that can create specificity for different substrates. This principle may be utilized by nature to achieve myriads of biological reactions using a limited set of amino acids and nucleotides. Sites composed of achiral moieties should be nonspecific with the interaction of moieties enclosed therein. Hence, active sites composed of chiral molecules can be used for carrying out diverse enantiospecific reactions.

1.3.2 *The influence of the confinement within a nanodimension on the reactions*

It was discussed earlier in the case of biomimetic systems that the intermolecular energy profile of a pair of chiral molecules at nanoscale separation has specific orientation-dependent minima, different for enantiomeric and racemic pairs. This difference is responsible for chiral discrimination, which vanishes at larger intermolecular separations. When the molecules are widely separated, the dissymmetric nature of the orientation-dependent interaction is absent. No preferential mutual orientation is observed over the mesoscopic length scale, and the manifestation of chiral discrimination vanishes. This is observed in the liquid-expanded and gaseous phases of the monolayer (Nandi and Vollhardt, 2003c, 4033; Nandi, 2009, 111). The same trend is noted in the case of a bilayer, where the helicity is observed at lower temperatures (Nandi and Bagchi, 1996, 11208; 1997, 1343). The reacting molecular segments in the active sites approach within nanoscale proximity in the course of a biological reaction. This proximity is important for chiral discrimination.

The active site residues not only enclose the reactants in a close separation but also confine them. This is a state of reduced degree of freedom for the reactants compared to the free state of the same moieties in the bulk, and this reduction in rotation and translation enhances the stereoselectivity. Due to the reduction in rotational and translational degrees of freedom, reactants with specific mutual orientations are more probable than random mutual arrangements. The orientation-dependent intermolecular interaction is more effective in the state of confinement in the active site, and this augments chiral discrimination in biological reactions. The restriction of degrees of freedom is an important factor in steric recognition. It has been long known that diffusion in reduced dimensions can speed up biological interaction over the limits normally set by three-dimensional diffusion processes (von Hippel and Berg, 1989, 675).

It has long been recognized that the enzymes, which are inherently chiral, are needed to orient both substrates and catalytic residues to preferentially stabilize one stereoisomeric transition state over another in the enzymatic reactions. The effectiveness of the catalytic activity of an enzyme in the rate enhancement significantly depends on the realization of tight binding, which is coupled with a high degree of steric recognition (Bruice, 1976, 331; Kirsch et al., 1984, 497). Proper orientation between the substrate and the catalytic residue is a necessary prerequisite for effective binding in the restriction of degrees of freedom. The fidelity of generation of L-configuration of the amino acid by transamination reaction is dependent on the addition of a proton to a particular face of the quinonoid intermediate, and the confinement and effective orientation play a role here. Successful mimics of catalytic enantioselective transamination fix the reactants in optimal geometry to enhance the reactivity (Bachman et al., 2004, 2044). A number of other examples are known similar to the transamination reaction, where the confinement within the nanodimension of the active site is important for stereoselectivity. For racemization and epimerization at the active sites of epimerase and racemase, it is necessary that the corresponding enzyme needs to clamp down on the substrate and should have been placed in a proximal relative position (Tanner, 2002, 237). Consequently, the confinement of the chiral moieties in a nanoscale separation is important for chiral discrimination (Nandi, 2009, 111).

In this book, some of the issues concerning the principles behind nature's technique in building the functional biological molecules, retention of their chiral structure, and replication of these structures through evolution as well as the accurate execution of the corresponding life processes have been discussed. The studies described in the book suggest that interactions of the active site residues with the substrate during the course of the reaction have commonality. For example, in the case of aminoacyl tRNA synthetase (aaRS), various interactions between the active site residues of a given aaRS, its cognate amino acid, and ATP can be categorized as substrate binding (ionic interaction, hydrogen bonding, and hydrophobic interaction), charge neutralization, and catalytic action. Some residues close to the reaction center can also have multiple roles. These functional roles are expected to be played by one or more active site residues of each of the different aaRSs during the course of the reaction. These functional residues are conserved in the active site and are distributed through regions of the active site surrounding the substrate moieties (ATP binding pocket, the amino acid binding pocket, and the reaction center in the case of aaRS) on the basis of their function. Few conserved active site residues play a more important role rather than merely anchoring the reactants at a suitable orientation in the reaction and are responsible for the stabilization of the transition state as well as retention of enantiopurity. While few residues have a dominant role such as catalytic activity, the active site as a whole (with its constituent amino acids that are close to the reactants) helps reduce the barrier height compared to the reaction in the absence of an active site. This understanding has emerged from a combination of structural analysis and computational understanding. It is worthwhile to explore the unifying principles underlying the preferential interaction and discrimination as exhibited by various enzymes in general.

One obvious utility of the related understanding is that it might be possible to design models of active sites using amino acids or nucleotides that can perform reactions involving new reactants with high fidelity which can act for the desired functionality for the reactants under consideration. Unless such a computational approach supplements the experimental approaches to designing a biocatalyst, the attempt will remain a random trial-and-error process rather than a logic-based design. Since enzyme engineering is dependent on the generation of variants of template enzymes and identification of most suited candidates, a quantitative understanding of the interactions is best suited for effective design. Combining these principles with the technological advancements already made might lead to the possible development of new synthetic architecture with desired functionality. The basic units of known materials of technological interests such as diverse polymers, crystals, liquid crystals, and gels are rather simple in structure. Biological functional architectures are much more sophisticated. Nature has always been superior in designing the world through molecules on which the very existence of life depends. The immensely efficient and orchestrated synthesis of myriad of biomolecules uses biophysical principles to achieve this. This book indicates that the confinement of the biomolecules in a nanoscale dimension and their chirality are utilized in biosynthetic reactions to enhance fidelity. Thus, gaining control over the methods of confinement of chiral moieties in a nanospace has vast potential for building newer functional structures with the desired physicochemical characteristics.

References

Andelman, D. (1989). Chiral discrimination and phase transitions in Langmuir monolayers. *J. Am. Chem. Soc.*, 111: 6536–6544.

Andelman, D. and Orland, H. (1993). Chiral discrimination in solutions and in Langmuir monolayers. *J. Am. Chem. Soc.*, 115: 12322–12329.

Armon, A., Graur, D., and Ben-Tal, N. (2001). ConSurf: An algorithmic tool for the identification of functional regions in proteins by surface mapping of phylogenetic information. *J. Mol. Biol.*, 307: 447–463.

Arnett, E.M., Harvey, N.G., and Rose, P.L. (1989). Stereochemistry and molecular recognition in two dimensions. *Acc. Chem. Res.*, 22: 131–148.

Arnez, J.G., Augustine, J.G., Moras, D., and Francklyn, C.S. (1997). The first step of aminoacylation at the atomic level in histidyl-tRNA synthetase. *Proc. Natl. Acad. Sci. (USA)*, 94: 7144–7149.

Avetisov, V. and Goldanskii, V. (1996). Mirror symmetry breaking at the molecular level. *Proc. Natl. Acad. Sci. (USA)*, 93: 11435–11442.

Bachman, S., Knudsen, K.R., and Jørgensen, K.A. (2004). Mimicking enzymatic transaminations: Attempts to understand and develop a catalytic asymmetric approach to chiral α-amino acids. *Org. Biomol. Chem.*, 2: 2044–2049.

Bartlett, G.J., Porter, C.T., Borkakoti, N., and Thornton, J.M. (2002). Analysis of catalytic residues in enzyme active sites. *J. Mol. Biol.*, 324: 105–121.

Benner S. (1993). Catalysis: Design versus selection. *Science*, 261: 1402–1403.

Benson, D.E., Wisz, M.S., and Hellinga, H.W. (2000). Rational design of nascent metalloenzymes. *Proc. Natl. Acad. Sci. (USA)*, 97: 6292–6297.

Berg, J.M., Tymoczko, J.L., and Stryer, L. (2002). *Biochemistry*. New York: W.H. Freeman.

Berova, N., Nakanishi, K., and Woody, R.W. (2000). *Circular Dichroism Principles and Applications*. New York: Wiley-VCH.

Berthier, D., Buffeteau, T., Léger, J.-M., Oda, R., and Huc, I. (2002). From chiral counterions to twisted membranes. *J. Am. Chem. Soc.*, 124: 13486–13494.

Bose, J.C. (1898). On the rotation of plane of polarisation of electric waves by a twisted structure. *Proc. Roy Soc. Lond. A*, 63: 146–152.

Bruice, T.C. (1976). Some pertinent aspects of mechanism as determined with small molecules. *Annu. Rev. Biochem.*, 45: 331–374.

Bryson, J.W., Betz, S.F., Lu, H.S., Suich, D.J., Zhou, H.X., O'Neil, K.T., and DeGrado, W.F. (1995). Protein design: A hierarchic approach. *Science*, 270: 935–941.

Buckingham, A.D. (2004). Chirality in NMR spectroscopy. *Chem. Phys. Lett.*, 398: 1–5.

Cantor, C.R. and Schimmel, P.R. (1980). *Biophysical Chemistry (Part I): The Conformation of Biological Macromolecules*. New York: W.H. Freeman.

Caparros, M., Torrecuadrada, J.L.M., and de Pedro, M.A. (1991). Effect of D-amino acids on *Escherichia coli* strains with impaired penicillin-binding proteins. *Res. Microbiol.*, 142: 345–350.

Caparros, M., Pisabarro, A.G., and de Pedro, M.A. (1992). Effect of D-amino acids on structure and synthesis of peptidoglycan in *Escherichia coli*. *J. Bacteriol.*, 174: 5549–5559.

Challis, G.L. and Naismith, J.H. (2004). Structural aspects of non-ribosomal peptide biosynthesis. *Curr. Opin. Str. Biol.*, 14: 748–756.

Champney, W.S. and Jensen, R.A. (1970). Molecular events in the growth inhibition of *bacillus subtilis* by D-tyrosine. *J. Bacteriol.*, 104: 107–116.

Chelliah, V., Chen, L., Blundell, T.L., and Lovell, S.C. (2004). Distinguishing structural and functional restraints in evolution in order to identify interaction sites. *J. Mol. Biol.*, 342: 1487–1504.

Cheng, G., Qian, B., Samudrala, R., and Baker, D. (2005). Improvement in protein functional site prediction by distinguishing structural and functional constraints on protein family evolution using computational design. *Nucleic Acids Res.*, 33: 5861–5867.

Chothia, C., Hubbard, T., Brenner, S., Barns, H., and Murzin, A. (1997). A. protein folds in the all-beta and all-alpha classes. *Annu. Rev. Biophys. Biomol. Struct.*, 26: 597–627.

Cosloy, S.D. and McFall, E. (1973). Metabolism of D-serine in *Escherichia coli* K-12: Mechanism of growth inhibition. *J. Bacteriol.*, 114: 685–694.

Craig, D.P., Power, E.A., and Thirunamachandran, T. (1971). The interaction of optically active molecules, *Proc. R. Soc. Lond. A*, 322: 165–179.

Dima, R.I. and Thirumalai, D. (2004). Asymmetry in the shapes of folded and denatured states of protein. *J. Phys. Chem. B*, 108: 6564–6570.

Dolain, C., Jiang, H., Léger, J.M., Guionneau, P., and Huc, I. (2005). Chiral induction in quinoline-derived oligoamide foldamers: Assignment of helical handedness and role of steric effects. *J. Am. Chem. Soc.*, 127: 12943–12951.

Elcock, A.H. (2001). Prediction of functionally important residues based solely on the computed energetics of protein structure. *J. Mol. Biol.*, 312: 885–896.

Elezzabi, A.Y. and Sederberg, S. (2009). Optical activity in an artificial chiral media: A terahertz time-domain investigation of Karl F. Lindman's 1920 pioneering experiment. *Opt. Express*, 17: 6600–6612.

Eliel, E.L., Wilen, S.H., and Mander, L.N. (1994). *Stereochemistry of Organic Compounds*. New York: John Wiley & Sons.

Fairbridges, R.W. and Jablonski, D. (eds) (1979). *The Encyclopaedia of Paleontology*. Stroudsburg, PA: Dowden, Hutchinson & Ross.

Flapan, E. (2000). *When Topology Meets Chemistry: A Topological Look at Molecular Chirality*. Cambridge, U.K.: Cambridge University Press.

Friedman, M. (1999). Chemistry, nutrition, and microbiology of D-amino acids. *J. Agric. Food Chem.*, 47: 3457–3479.

Fuhrhop, J.H. and Helfrich, W. (1993). Fluid and solid fibers made of lipid molecular bilayers. *Chem. Rev.*, 93: 1565–1582.

Fuji, N. (2002). D-amino acids in living higher organisms. *Orig. Life Evol. Biosph.*, 32: 103–127.

Fujiki, M., Nakashima, H., Toyoda, S., and Koe, J.R. (2003). Chirality in polysilanes. In Green, M., Nolte, R.J.M., Meijer, E.W. (eds), *Materials Chirality: Volume 24 of Topics in Stereochemistry Series*, and Denmark, S.E., Siegel, J. (series eds). New York: Wiley Interscience, John Wiley & Sons, 209–279.

Gerlt, J.A. (1987). Relationships between enzymatic catalysis and active site structure revealed by applications of site-directed mutagenesis. *Chem. Rev.*, 87: 1079–1105.

Green, M.M., Cheon, K., Yang, S., Park, J., Swansburg, S., and W. Liu. (2001). Chiral studies across the spectrum of polymer science. *Acc. Chem. Res.*, 34: 672–680.

Guerra, G., Cavallo, L., and Corradini, P. (2003). Chirality of catalysts for stereospecific polymerizations. in Green, M., Nolte, R.J.M., and Meijer, E.W. (eds), *Materials Chirality: Volume 24 of Topics in Stereochemistry Series*, and Denmark, S. E., Siegel, J. (series eds). New York: Wiley Interscience, John Wiley & Sons, 1–69.

Gunde, K.E., Gary, W., Burdicka, G.W., and Richardson, S. (1996). Chirality-dependent two-photon absorption probabilities and circular dichroic line strengths: Theory, calculation and measurement. *Chem. Phys.*, 208: 195–219.

Gutteridge, A., Bartlett, G.J., and Thornton, J.M. (2003). Using a neural network and spatial clustering to predict the location of active sites in enzymes. *J. Mol. Biol.*, 330: 719–734.

Handel, T., Williams, S., and DeGrado W. (1993). Metal ion-dependent modulation of the dynamics of a designed protein. *Science*, 261: 879–885.

Hansen, J.L., Schmeing, M.T., Moore, P.B ., and Steitz, T.A. (2002). Structural insights into peptide bond formation. *Proc. Natl. Acad. Sci.* (USA), 99: 11670–11675.

Harada, N. and Nakanishi, K. (1972). The exciton chirality method and its application to configurational and conformational studies of natural products. *Acc. Chem. Res.*, 5: 257–263.

Harris, C.L. (1981). Cysteine and growth inhibition of *Escherichia coli*: Threonine deaminase as the target enzyme. *J. Bacteriol.*, 145: 1031–1035.

Hegstrom, R. and Kondepudi, D.K. (1990). The handedness of the universe. *Sci. Am.*, 262: 108–115.

Horsman, G.P., Liu, A.M.F., Henke, E., Bornscheuer, U.T., and Kazlauskas, R.J. (2003). Mutations in distant residues moderately increase the enantioselectivity of *Pseudomonas fluorescens esterase* towards methyl 3-bromo-2-methylpropanoate and ethyl 3-phenylbutyrate. *Chem. Eur. J.*, 9: 1933–1939.

Humphrey, W., Dalke, A., and Schulten, K. (1996). VMD: Visual molecular dynamics. *J. Mol. Graphics*, 14: 33–38.

Jacques, J., Collet, A., and Wilen, S.H. (1981). *Enantiomers, Racemates, and Resolutions*. New York: John Wiley & Sons.

Janoschek, R. (ed) (1991). *Chirality—From Weak Boson to the Alpha-Helix*. New York: Springer-Verlag.

Jiang, H., Dolain, C., Léger, J.M., Gornitzka, H., and Huc, I. (2004). Switching of chiral induction in helical aromatic oligoamides using solid state–solution state equilibrium. *J. Am. Chem. Soc.*, 126: 1034–1035.

Jones, S. and Thornton, J.M. (2004). Searching for functional sites in protein structures. *Curr. Opin. Chem. Biol.*, 8: 3–7.

Joyce, G.F. and Orgel, L.E. (1999). Prospects for understanding the origin of the RNA world. In Gesteland, R.F., Cech, T.R., Atkins, J.F. (eds). *The RNA World*, 2nd ed., Cold Spring Harbor, NY: Cold Spring Harbor Laboratory Press, 49–77.

Kamtekar, S., Schiffer, J., Xiong, H., Babik, J., and Hecht, M. (1993). Protein design by binary patterning of polar and nonpolar amino acids. *Science*, 262: 1680–1685.

Kirsch, J.F., Eichele, G., Ford, G.C., Vincent, M.G., Jansonius, J.N., Gehring, H., and Christen, P. (1984). Mechanism of action of aspartate aminotransferase proposed on the basis of its spatial structure. *J. Mol. Biol.*, 174: 497–525.

Knobler, C.M. (1990). Recent developments in the study of monolayers at the air-water interface. *Adv. Chem. Phys.*, 77: 397–425.

Kondepudi, D.K. and Durand, D.J. (2001). Chiral asymmetry in spiral galaxies? *Chirality*, 13: 351–356.

Kortemme, T. and Baker, D. (2002). A simple physical model for binding energy hot spots in protein–protein complexes. *Proc. Natl. Acad. Sci. (USA)*, 99: 14116–14121.

Kreil, G. (1997). D-amino acids in animal peptides. *Annu. Rev. Biochem.*, 66: 337–345.

Kuroda, R., Mason S.F., Rodger, C.D., and Seal R.H. (1981). Chiral discrimination—an extended transition monopole model. *Mol. Phys.*, 42: 33–50.

Kuroda, R., Mason S.F., Rodger, C.D., and Seal, R.H. (1978). A stereochemical basis for chiral discrimination: The extended transition monopole model. *Chem. Phys. Lett.*, 57: 1–4.

Landgraf, R., Xenarios, I., and Eisenberg, D. (2001). Three-dimensional cluster analysis identifies interfaces and functional residue clusters in proteins. *J. Mol. Biol.*, 307: 1487–1502.

Laskowski, R.A., Luscombe, N.M., Swindells, M.B., and Thornton, J.M. (1996). Protein clefts in molecular recognition and function. *Protein Sci.*, 5: 2438–2452.

Leitereg, T.J., Guadagni, D.G., Harris, J., Mon, T.R., and Teranishi, R. (1971). Evidence for the difference between the odours of the optical isomers (+)- and (−)-carvone. *Nature*, 230: 455–456.

Leporl, L., Mengheri, M., and Mollica, V. (1983). Discriminating interactions between chiral molecules in the liquid phase: Effect on volumetric introduction properties. *J. Phys. Chem.*, 87: 3520–3525.

Lichtarge, O., Bourne, H.R., and Cohen, F.E. (1996). An evolutionary trace method defines binding surfaces common to protein families. *J. Mol. Biol.*, 257: 342–358.

Madabushi, S., Yao, H., Marsh, M., Kristensen, D.M., Philippi, A., Sowa, M.E., and Lichtarge, O. (2002). Structural clusters of evolutionary trace residues are statistically significant and common in proteins. *J. Mol. Biol.*, 316: 139–154.

Marahiel, M.A., Stachelhaus, T., and Mootz, H.D. (1997). Modular peptide synthetases involved in nonribosomal peptide synthesis. *Chem. Rev.*, 97: 2651–2674.

Marti, S., Andres, J., Moliner, V., Silla, E., Tunon, I., and Bertran, J. (2008). Computational design of biological catalysts. *Chem. Soc. Rev.*, 37: 2634–2643.

Mason, S. (1988). Biomolecular homochirality. *Chem. Soc. Rev.*, 17: 347–359.

Mason, S.F. (1982). *Molecular Optical Activity and the Chiral Discriminations.* Cambridge, U.K.: Cambridge University Press.

Mason, S.F. (1989). The development of concepts of chiral discrimination. *Chirality*, 1: 183–191.

Mason, S.F. and Tranter, G.E. (1985). The electroweak origin of biomolecular handedness. *Proc. Roy. Soc. London A*, 397: 45–65.

McConnell, H.M. (1991). Structures and transitions in lipid monolayers at the air/water interface. *Annu. Rev. Phys. Chem.*, 42: 171–195.

McGhee, G.R. (1989). In Donovan S.K. Jr. (ed). *Mass Extinctions: Process and Evidence.* London: Belhaven Press, 133.

Möhwald, H. (1990). Phospholipid and phospholipid-protein monolayers at the air/water interface. *Annu. Rev. Phys. Chem.*, 41: 441–476.

Morley, K.L. and Kazlauskas, R.J. (2005). Improving enzyme properties: When are closer mutations better. *Trends Biotechnol.*, 23: 231–237.

Moss, G.P. Recommendations of the Nomenclature Committee. International Union of Biochemistry and Molecular Biology on the Nomenclature and Classification of Enzymes by the Reactions They Catalyze. <http://www. chem.qmul.ac.uk/iubmb/enzyme/. Retrieved 2006-03-14.

Nandi, N. (2003). Molecular origin of the recognition of chiral odorant by chiral lipid: Interaction of dipalmitoyl phosphatidyl choline and carvone. *J. Phys. Chem. A*, 107: 1588–1591.

Nandi, N. (2004). Role of secondary level chiral structure in the process of molecular recognition of ligand: Study of model helical peptide. *J. Phys. Chem. B.*, 108: 789–797.

Nandi, N. (2009). Chiral discrimination in the confined environment of biological nanospace: Reactions and interactions involving amino acids and peptides. *Int. Rev. Phys. Chem.*, 28: 111–167.

Nandi, N. and Bagchi, B. (1996). Molecular origin of the intrinsic bending force for helical morphology observed in chiral amphiphilic assemblies: Concentration and size dependence. *J. Am. Chem. Soc.*, 118: 11208–11216.

Nandi, N. and Bagchi, B. (1997). Prediction of the senses of helical amphiphilic assemblies from effective intermolecular pair potential: Studies on chiral monolayers and bilayers. *J. Phys. Chem. A.*, 101: 1343–1351.

Nandi, N. and Vollhardt, D. (2000). Microscopic study of chiral interactions in Langmuir monolayer: Monolayers of *N*-palmitoyl aspartic acid and *N*-stearoyl serine methyl ester. *Colloids Surf. A*, 183–185: 67–83.

Nandi, N. and Vollhardt, D. (2002a). Molecular origin of the chiral interaction in biomimetic systems: Dipalmitoylphosphatidylcholine Langmuir monolayer. *J. Phys. Chem. B.*, 106: 10144–10149.

Nandi, N. and Vollhardt, D. (2002b). Prediction of the handedness of the chiral domains of amphiphilic monolayers: Monolayers of amino acid amphiphiles. *Colloids Surf. A.*, 198–200: 207–221.

Nandi, N. and Vollhardt, D. (2003a). Chiral discrimination effects in Langmuir monolayers: Monolayers of palmitoyl aspartic acid, n-stearoyl serine methyl ester, and n-tetradecyl-γ,δ dihydroxypentanoic acid amide. *J. Phys. Chem. B*, 107: 3464–3475.

Nandi, N. and Vollhardt, D. (2003b). Correlation between the microscopic and mesoscopic chirality in Langmuir monolayers. *Thin Solid Films*, 433: 12–21.

Nandi, N. and Vollhardt, D. (2003c). Effect of molecular chirality on the morphology of biomimetic Langmuir monolayers. *Chem. Rev.*, 103: 4033–4075.

Nandi, N. and Vollhardt, D. (2006). Chirality and molecular recognition in biomimetic organized films. In Ariga, K. and Nalwa, H.S. (eds), *Bottoms up Nanofabrication: Supramolecules, Self Assemblies and Organized Films*. Valencia, CA: American Scientific Publishers, 5: 1–29.

Nandi, N. and Vollhardt, D. (2007). Molecular interactions in amphiphilic assemblies: Theoretical perspective. *Acc. Chem. Res.*, 40: 351–360.

Nandi, N. and Vollhardt, D. (2008). Chiral discrimination and recognition in Langmuir monolayers. *Curr. Opin. Colloid Surf.*, 13: 40–46.

Nandi, N., Roy, R.K., Anupriya, Upadhaya, S., and Vollhardt, D. (2002). Chiral interaction in enantiomeric and racemic dipalmitoyl phosphatidylcholine Langmuir monolayer. *J. Surf. Sci. Technol.*, 18: 51–66.

Nandi, N., Thirumoorthy, K., and Vollhardt, D. (2008). Chiral discrimination in stearoyl amine glycerol monolayers. *Langmuir*, 24: 9489–9494.

Nandi, N., Vollhardt, D., and Brezesinski, G. (2004). Chiral discrimination effects in Langmuir monolayers of 1-O-hexadecyl glycerol. *J. Phys. Chem. B*, 108: 327–335.

Nandi, N., Vollhardt, D., and Rudert, R. (2004). Molecular pair potential of chiral amino acid amphiphile in Langmuir monolayers on the basis of an atomistic model. *Colloids Surf. A*, 250: 279–287.

Novotny, M. and Kleywegt, G.J. (2005). A survey of left-handed helices in protein structures. *J. Mol. Biol.*, 347: 231–241.

Ohnishi, E., Macleod, H., and Horowitz, N.H. (1962). Mutants of neurospora deficient in D-amino acid oxidase. *J. Biol. Chem.*, 237: 138–142.

Okamoto, Y., Yashima, E., and Yamamoto, C. (2003). Optically active polymers with chiral recognition ability. In Green, M., Nolte, R.J.M., Meijer, E.W. (eds), *Materials Chirality: Volume 24 of Topics in Stereochemistry Series*, and Denmark, S.E., Siegel, J. (series eds). New York: Wiley Interscience, John Wiley & Sons, 157–207.

Ondrechen, M.J., Clifton, J.G., and Ringe, D. (2001). Thematics: A simple computational predictor of enzyme function from structure. *Proc. Natl. Acad. Sci. (USA)*, 98: 12473–12478.

Otten, L.G., Hollmann, F., and Arends, I.W.C.E. (2010). Enzyme engineering for enantioselectivity: From trial-and-error to rational design? *Trends Biotechnol.*, 28: 46–54.

Oue, S., Okamoto, A., Yano, T., and Kagamiyama, H. (1999). Redesigning the substrate specificity of an enzyme by cumulative effects of the mutations of non-active site residues. *J. Biol. Chem.*, 274: 2344–2349.

Pályi, G., Zucchi, C., and Caglioti, L. (1999). In Pályi, G., Zuccchi, C., Caglioti, L. (eds), Dimensions of biological homochirality. *Advances in Biochirality*. Oxford: Elsevier Science S.A., 3–12.

Pei, J., Dokholyan, N.V., Shakhnovich, E.I., and Grishin, N.V. (2003). Using protein design for homology detection and active site searches. *Proc. Natl. Acad. Sci. (USA)*, 100: 11361–11366.

Pérez-Payá, E., Houghten, R.A., and Blondelle S.E. (1996). Functionalized protein-like structures from conformationally defined synthetic combinatorial libraries. *J. Biol. Chem.*, 271: 4120–4126.

Porter, C.T., Bartlett, G.J., and Thornton, J.M. (2004). The catalytic site atlas: A resource of catalytic sites and residues identified in enzymes using structural data. *Nucleic Acids Res.*, 32: D129–D133.

Rosa, C.D. (2003). Chain conformation, crystal structures, and structural disorder in stereoregular polymers. In Green, M., Nolte, R.J.M., Meijer, E.W. (eds), *Materials Chirality: Volume 24 of Topics in Stereochemistry Series*, and Denmark, S.E., Siegel, J. (series eds). New York: Wiley Interscience, John Wiley & Sons, 71–155.

Rytka, J. (1975). Positive selection of general amino acid permease mutants in *Saccharomyces cerevisiae*. *J. Bacteriol.*, 121: 562–570.

Salem, L., Chapuisat, X., Segal, G., Hiberty, P.C., Minot, C., Leforesteir, C., and Sautet, P. (1987). Chirality forces. *J. Am. Chem. Soc.*, 109: 2887–2894.

Sandars, P.G.H. (2005). Chirality in the RNA world and beyond. *Int. J. Astrobiology*, 4: 49–61.

Schneider, T.D. (1997). Information content of individual genetic sequences. *J. Theor. Biol.*, 189: 427–441.

Schnur, J.M. (1993). Lipid tubules: A paradigm for molecularly engineered structures. *Science*, 262: 1669–1676.

Selinger, J.V., Spector, M.S., and Schnur, J.M. (2001). Theory of self-assembled tubules and helical ribbons. *J. Phys. Chem. B.*, 105: 7157–7169.

Shallenberger, R.S., Acree, T.E., and Lee, C.Y. (1969). Sweet taste of D and L-sugars and amino-acids and the steric nature of their chemo-receptor site. *Nature*, 221: 555–556.

Soai, K., Shibata, T., Morioka, H., and Choji, K. (1995). Asymmetric autocatalysis and amplification of enantiomeric excess of a chiral molecule. *Nature*, 378: 767–768.

Stachelhaus, T. and Walsh, C.T. (2000). Mutational analysis of the epimerization domain in the initiation module PheATE of gramicidin S synthetase. *Biochemistry*, 39: 5775–5787.

Taniguchi, T., Martin, C.L., Monde, K., Nakanishi, K., Berova, N., and Overman, L.E. (2009). Absolute configuration of actinophyllic acid as determined through chiroptical data. *J. Nat. Prod.*, 72: 430–432.

Tanner, M.E. (2002). Understanding nature's strategies for enzyme-catalyzed racemization and epimerization. *Acc. Chem. Res.*, 35: 237–246.

Taylor, T.N. and Taylor, E.L. (1993). *The Biology and Evolution of Fossil Plants.* Englewood Cliffs, NJ: Prentice Hall, 118.

Taylor, W.R. (1986). The classification of amino acid conservation. *J. Theor. Biol.*, 119: 205–218.

Thirumoorthy, K. and Nandi, N. (2005). The correlation between the molecular chirality of the sugar ring on the mesoscopic aggregate morphology in RNA and DNA mimetic systems. *Chem. Phys. Lett.*, 414: 336–430.

Thirumoorthy, K. and Nandi, N. (2006). Comparison of the intermolecular energy surfaces of amino acids: Orientation-dependent chiral discrimination. *J. Phys. Chem. B*, 110: 8840–8849.

Thirumoorthy, K., Nandi, N., and Vollhardt, D. (2006). Prediction of the handedness of the domains of monolayers of D-N-palmitoyl aspartic acid: Integrated molecular orbital and molecular mechanics based calculation. *Colloids Surf. A*, 282–283: 222–226.

Todd, A.E., Orengo, C.A., and Thornton, J.M. (2001). Evolution of function in protein superfamilies from a structural perspective. *J. Mol. Biol.*, 307: 1113–1143.

Torrance, J.W., Bartlett, G.J., Porter, C.T., and Thornton, J.M. (2005). Using a library of structural templates to recognize catalytic sites and explore their evolution in homologous families. *J. Mol. Biol.*, 347: 565–581.

Torres, A.M., Tsampazi, M., Kennett, E.C., Belov, K., Geraghty, D.P., Bansal, P.S., Alewood, P.F., and Kuchel, P.W. (2007). Characterization and isolation of L-to-D-amino-acid-residue isomerase from platypus venom. *Amino Acids*, 32: 63–68.

Tsuruoka, T., Tamura, A., Miyata, A., Takei, T., Iwamatsu, K., Inouye, S., and Matsuhashi, M. (1984). Penicillin-insensitive incorporation of D-amino acids into cell wall peptidoglycan influences the amount of bound lipoprotein in *Escherichia coli*. *J. Bacteriol.*, 160: 889–894.

Vollhardt, D. (1996). Morphology and phase behavior of monolayers. *Adv. Colloid Interface Sci.*, 64: 143–171.

von Döhren, H., Keller, U., Vater, J., and Zocher, R. (1997). Multifunctional peptide synthetases. *Chem. Rev.*, 97: 2675–2705.

von Hippel P.H. and Berg, O.G. (1989). Facilitated target location in biological system. *J. Biol. Chem.*, 264: 675–678.

Wang, K. and Samudrala, R. (2005). FSSA: A novel method for identifying functional signatures from structural alignments. *Bioinformatics*, 21: 2969–2977.

Warshel, A. (1978). Energetics of enzyme catalysis. *Proc. Natl. Acad. Sci. (USA)*, 75: 5250–5254.

Warshel, A., Sharma, P.K., Kato, M., Xiang, Y., Liu, H., and Olsson, M.H.M. (2006). Electrostatic basis for enzyme catalysis. *Chem. Rev.*, 106, 3210–3235.

Webb, Edwin C. (1992). Enzyme nomenclature 1992: Recommendations of the Nomenclature Committee of the International Union of Biochemistry and Molecular Biology on the Nomenclature and Classification of Enzymes. San Diego: Published for the International Union of Biochemistry and Molecular Biology by Academic Press. ISBN 0-12-227164-5. http://www.chem.qmul.ac.uk/iubmb/enzyme/.

Wenzel, T.J. (2007). *Discrimination of Chiral Compounds Using NMR Spectroscopy.* Hoboken, New Jersey: John Wiley & Sons.

Xiang, H., Luo, L., Taylor, K.L., and Dunaway-Mariano, D. (1999). Interchange of catalytic activity within the 2-enoyl-coenzyme A hydratase/isomerase superfamily based on a common active site template. *Biochemistry*, 38: 7638–7652.

Youngman, E.M., Brunelle, J.L., Kochaniak, A.B., and Green, R. (2004). The active site of the ribosome is composed of two layers of conserved nucleotides with distinct roles in peptide bond formation and peptide release. *Cell*, 117: 589–599.

chapter two

Chiral discrimination in the active site of oxidoreductases

Oxidoreductases are enzymes that catalyze oxidation-reduction reactions. The accepted name could be donor-dehydrogenase. The term *oxidase* is used only when oxygen is an acceptor. The active sites of a number of oxidoreductases are studied, which reveals that the binding of different molecules such as drugs could be chirality specific.

2.1 Cytochrome P450: discrimination in drug (warfarin) interaction

The cytochrome P450 superfamily (abbreviated as CYP) is a large and diverse group of enzymes that catalyzes oxidation of organic substances. CYPs are often part of multicomponent electron transfer chains that are called P450-containing systems. CYPs catalyze a variety of reactions, for example, a monooxygenase reaction that involves insertion of one atom of oxygen into an organic substrate while the other oxygen atom is reduced to water (after the product is released from the active site). The enzyme returns to its original state, and simultaneously, a water molecule returns to occupy the distal coordination position of the iron nucleus. Cytochrome P450 proteins (CYP450s) metabolize physiologically important compounds in many species of microorganisms, plants, and animals. Human CYP450 proteins such as CYP1A2, CYP2C9, CYP2C19, CYP2D6, and CYP3A4 are the major drug-metabolizing isoforms. These CYPs carry out the oxidative metabolism of more than 90% of the drugs in current clinical use. A detailed list of substrates and inhibitors of CYP1A2, CYP2B6, CYP2C8, CYP2C9, CYP2C19, CYP2D6, CYP2E1, CYP3A4, CYP3A5, and CYP3A7 is available in the literature (Flockhart, 2007). The variety of the drugs indicates the diverse functionality of the CYP enzymes. Polymorphic variants have also been reported for some CYP450 isoforms, which has implications for the efficacy of drugs in individuals, and for the coadministration of drugs (Williams et al., 2003, 464). Notably, many drugs are chiral, but the molecular basis of drug recognition by human CYP450s, however, has remained elusive. The catalytic activity of eukaryotic P450 shows regioselectivities and also enantiospecificities toward many substrates.

Based on the crystal structure of bacterial P450 101 (Poulos et al., 1985, 16122; 1987, 687; Raag et al., 1991, 11420) and to alignments of P450 amino acid sequences (Gotoh, 1992, 83), different important features of the eukaryotic P450 enzyme are now understood. Protein engineering studies of bacterial P450 101 (Atkins and Sligar, 1988, 18842; 1989, 8742; Gerber and Sligar, 1992, 8742; Imai and Nakamura, 1989, 717; Imai et al., 1989, 7823; Martinis et al., 1989, 9252) and eukaryotic P450s (Furuya et al., 1989, 6848; Graham-Lorence et al., 1991, 11939; Hanioka et al., 1992, 3364; Ishigooka et al., 1992, 1528; Iwasaki et al., 1991, 3380; Zhou et al., 1992, 762) indicate that amino acids at the (putative) distal site also contribute to the catalytic activities of this enzyme. The effect may be due to a couple of factors. First, the distal site contains a relatively greater number of residues compared to the fewer number of residues close to the active site. Second, due to the long-range nature of the electrostatic interaction of the distal site residues with reactants, it is not surprising that their influence is nonnegligible.

The active site of cytochrome P450 contains a heme iron center. The iron is tethered to the P450 protein via a thiolate ligand derived from a cysteine residue. This cysteine and several surrounding residues are highly conserved in known CYPs and have the formal PROSITE signature consensus pattern [FW]-[SGNH] - x - [GD] -{F}- [RKHPT] - {P}-C-[LIVMFAP]-[GAD], where C is the heme iron ligand. P450 IIB10 from mouse has Lys in the first position of the pattern, and is an exception. The bound substrate induces a change in the conformation of the active site, often displacing a water molecule from the distal axial coordination position of the heme iron (Meunier et al., 2004, 3947). The change in the electronic state of the active site favors the transfer of an electron from nicotinamide adenine dinucleotide phosphate (NADPH) via cytochrome P450 reductase or another associated reductase. This takes place by way of the electron transfer chain, as described earlier, reducing the ferric heme iron to the ferrous state.

Detailed analysis of the binding sites of the CYP2C subfamily has been carried out (Meng et al., 2009, 1066). A comparative analysis of the binding sites of mouse CYP2C38 and CYP2C39 based on homology modeling, molecular dynamics simulation, and docking studies has been done. While S-warfarin is the substrate for CYP2C9, R-warfarin is the substrate for CYP1A2 and CYP2C19. The unliganded crystal structures of CYP2C9 (a human CYP450) and as a complex with the anticoagulant drug warfarin (S-warfarin) have been studied (Williams et al., 2003, 464). The chemical structure of warfarin is shown in Figure 2.1. Warfarin is a derivative of coumarin and can prevent thrombosis and embolism (abnormal formation and migration of blood clots). This compound was initially marketed as a pesticide against rats and mice and is being currently used for this purpose, but it has also been found to be effective in many disorders.

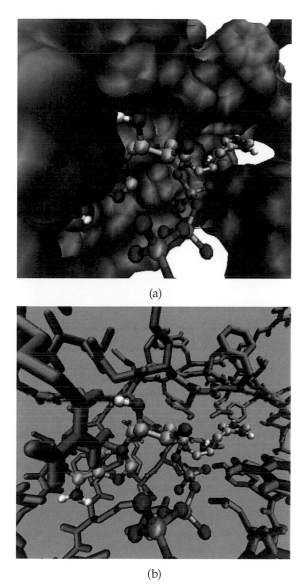

(a)

(b)

Figure 1.7 (a) Schematic representation of the active site of prokaryotic *Escherichia coli* Histidyl tRNA synthetase generated from the crystal structures 1KMM.PDB and 1KMN.PDB (Arnez et al., 1997, 7144). The substrates (histidinol and ATP) are shown by a ball and stick representation and the enclosing nanospace (cavity of the active site) is shown as a gray surface. (b) Schematic representation of the active site of prokaryotic *Escherichia coli* (same as in Figure 1.7a) in which the amino acids constituting the enclosing nanospace (cavity of the active site) are explicitly shown with the stick representation. The images were prepared using VMD (Humphrey, W. et al., 1996. *J. Mol. Graphics*, 14: 33).

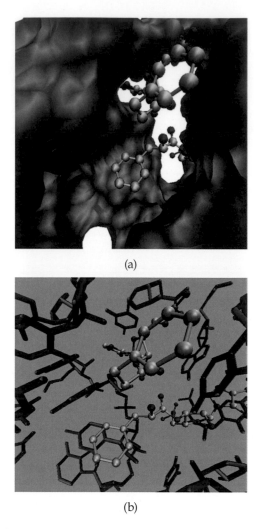

(a)

(b)

Figure 1.8 (a) Schematic representation of the active site of architecture of the peptidyl transferase center (PTC) taken from the crystal structure of the ribosomal part of *Haloarcula marismortui* (Hansen et al., 2002, 11670). The crystal structure of CCA-Phe-cap-biotin bound simultaneously at half-occupancy to both the A-site and P-site of the 50S ribosomal subunit (1Q86.PDB) is used to generate the structure. The reactants (amino (A) and peptidyl (P) terminals) are shown by a ball and stick representation and the enclosing nanospace (cavity of the active site) is shown as a gray surface. (b) Schematic representation of the active site of architecture of the peptidyl transferase center (PTC) (same as in Figure 1.8a) in which the nucleotides constituting the enclosing nanospace (cavity of the active site) are explicitly shown with the stick representation. The images were prepared using VMD (Humphrey, W. et al., 1996. *J. Mol. Graphics*, 14: 33).

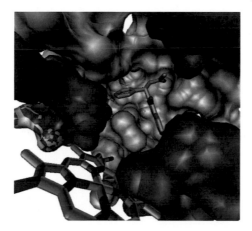

Figure 2.2 The structure of S-warfarin bound with the CYP450 2C9. The heme iron center is far from the warfarin (1OG2.PDB). The image was generated using VMD (Humphrey, W. et al., 1996. *J. Mol. Graphics*, 14: 33).

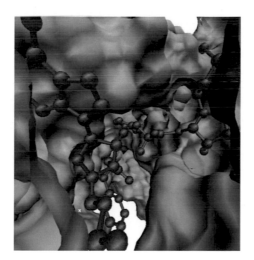

Figure 3.2 Schematic representation of the active site of the architecture of peptidyl transferase center (PTC) taken from the crystal structure of the ribosomal part of *Haloarcula marismortui* (Hansen, J.L. et al., 2002. *Proc. Natl. Acad. Sci. (USA)*. 99: 11670). The crystal structure of CCA-Phe-cap-biotin bound simultaneously at half-occupancy to both the A-site and P-site of the 50S ribosomal subunit (1Q86. PDB) is used to generate the structure. The reactants (amino (A) and peptidyl (P) terminals) are shown by ball and stick representation, and the enclosing nano-space (cavity of the active site) is shown as a green surface.

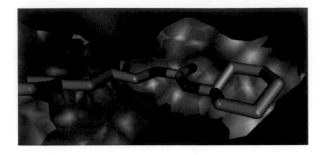

Figure 4.1 Active site of liver cytosolic epoxide hydrolase (1CR6.PDB). The image is generated using VMD (Humphrey et al., 1996: 33).

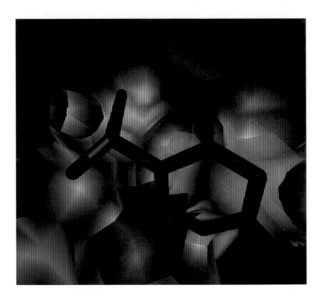

Figure 5.2 The view of the active site and the channel linking to the protein surface of the hydroxynitrile lyase (1YB6.PDB) from the tropical rubber tree *Hevea brasiliensis* (Gartler, G. et al., 2007. *J. Biotechnol.*, 129: 87).

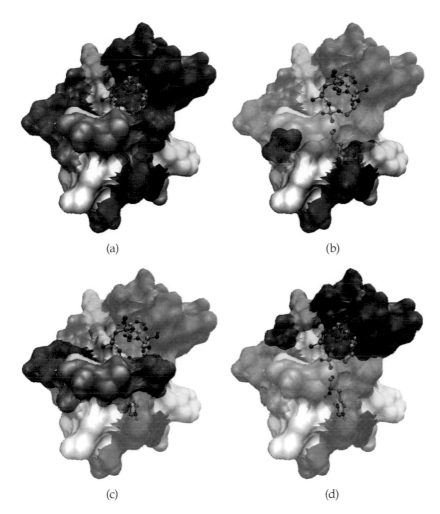

(a)   (b)

(c)   (d)

Figure 6.1 Schematic representation of the (a) active site of prokaryotic *Escherichia coli* histidyl-tRNA synthetase (1KMN.PDB), (Arnez et al., 1997. *Proc. Natl. Acad. Sci. (USA)*, 94: 7144) which includes the amino acid binding site, ATP binding site, and intervening region; (b) amino acid binding pocket; (c) intervening region; and (d) ATP binding pocket. The reactants (histidinol, an inhibitor and ATP) are shown by ball and stick representation, and the surrounding active site cavity is shown by surface representation. The images are prepared using VMD (Humphrey et al., 1996: 33).

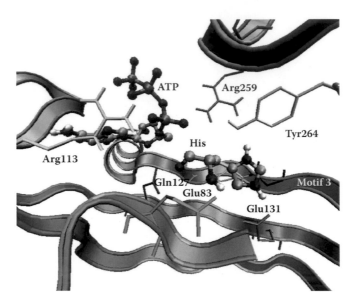

Figure 6.3 Active site of the HisRS-Histidinol-ATP complex (1KMN.PDB) from *E. coli*. The reactants (histidinol and ATP) are shown in CPK style. Side chains of the active site in close proximity to the reactants as Glu 83, Arg 113, Gln 127, Glu 131, Arg 259, and Tyr 264 are shown by bonds. The remaining part of the motif 2, His A loop, and motif 3 are represented by ribbons. (Dutta Banik, S. and Nandi, N. 2010. *J. Phys. Chem. B.*, 114: 2301.) The image is prepared using VMD. (Reprinted with permission from the American Chemical Society. © American Chemical Society.)

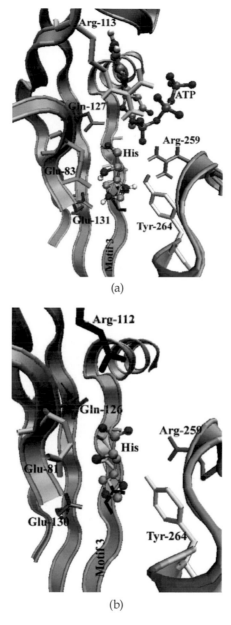

(a)

(b)

Figure 6.11 (a) The representation of the active site of three prokaryotic HisRS *E. coli* (b), *TT,* and (c) *SA* as obtained from the respective crystal structure. The images are prepared using VMD (Humphrey et al., 1996: 33).   (continued)

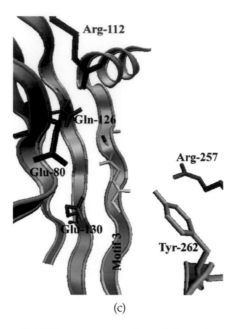

(c)

Figure 6.11 (continued) (a) The representation of the active site of three prokaryotic HisRS *E. coli* (b), *TT,* and (c) *SA* as obtained from the respective crystal structure. The images are prepared using VMD (Humphrey et al., 1996: 33).

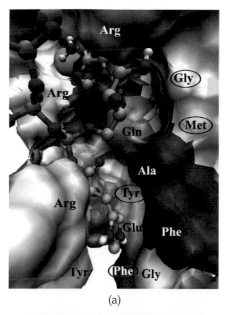

(a)

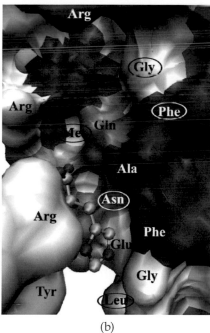

(b)

Figure 6.12 Comparison of the active site residues in HisRS of three prokaryotic species, *E. coli*, *TT*, and *SA*. The residues which are nonconserved in three prokaryotic HisRS (a) *E. coli* and (b) *TT* are encircled in the figure. The images are prepared using VMD (Humphrey et al., 1996: 33).   (continued)

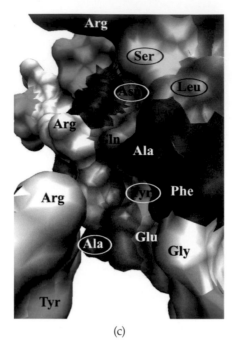

(c)

Figure 6.12 (continued) Comparison of the active site residues in HisRS of three prokaryotic species, *E. coli*, *TT*, and *SA*. The residues which are nonconserved in three prokaryotic HisRS (c) *SA* are encircled in the figure. The images are prepared using VMD (Humphrey et al., 1996: 33).

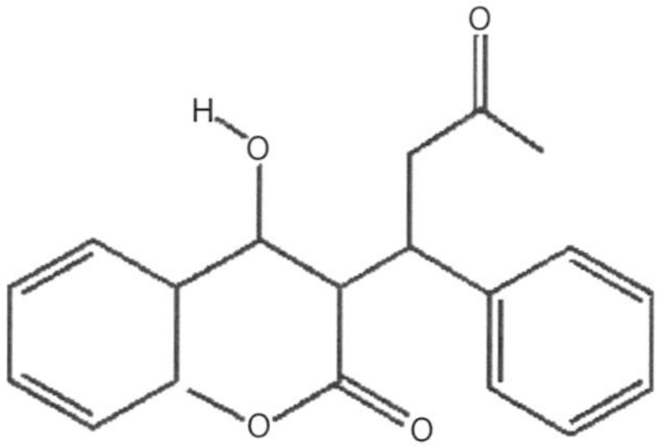

Figure 2.1 The chemical structure of warfarin ((RS)-4-hydroxy-3-(3-oxo-1-phenyl-butyl)-2H-chromen-2-one).

CYP2C9 catalyzes the 6- and 7-hydroxylation of the active enantiomer of warfarin (S-warfarin) to inactive metabolites. The structure shows interactions between CYP2C9 and warfarin, and reveals a new binding pocket. The binding mode of warfarin suggests that CYP2C9 may undergo an allosteric mechanism during its function. It is suggested that the active site of CYP2C9 may simultaneously accommodate multiple ligands during its biological function. S-warfarin lies in a predominantly hydrophobic pocket lined by residues such as Arg, Gly, Ile, Leu, Ala, Val, Asn, Thr, Ser, Leu, Pro, and multiple Phe. More specifically, the phenyl group of S-warfarin packs against the side chains of Phe476 and Phe100 and also contacts Pro367. Although the binding of S-warfarin to CYP2C9 appears not to induce major conformational changes within the protein, some local rearrangements are observed. The side chain of Phe476, which showed conformational mobility in the substrate-free structure, forms a π- π stacking interaction with the S-warfarin phenyl group.

The structure of S-warfarin bound with the CYP450 2C9 is shown in Figure 2.2. The bicyclic scaffold of S-warfarin makes van der Waals contact with the side chains of Ala103, Phe114, and Pro367. Hydrogen-bonding interactions are observed between carbonyl oxygen atoms of the warfarin and backbone amide nitrogen atoms of Phe100 and Ala103. The 4-hydroxycoumarin ring of the S-warfarin interacts with the Ile99 and Phe100 (via main chain) as well as Ala103 and Pro367 (via side chain). The keto group of 4-hydroxycoumarin forms a hydrogen bond with the amide group of Phe100. The phenyl group attached with the chiral center of warfarin interacts with the Phe476. The oxo group of the 3-oxo-phenylbutyl group forms a hydrogen bond with the amide hydrogen atom of Ala103. The network interactions stabilize the S-warfarin into the binding site located far from the heme group and form the basis of preferential interaction with the S-chirality of warfarin. The residues close to warfarin are shown in Figure 2.3. It is proposed that the compound moves from this primary recognition site toward the heme to facilitate catalysis.

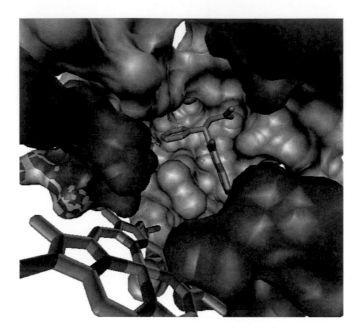

Figure 2.2 **See color insert.** The structure of S-warfarin bound with the CYP450 2C9 (1OG2.PBD). The heme iron center is far from the warfarin. The image was generated using VMD (Humphrey, W. et al., 1996. *J. Mol. Graphics*, 14: 33).

It is suggested that such a two-step conformational movement may be triggered by an electron-transfer-driven conformational change within CYP2C9. Clearly, the orientation and the binding of S-warfarin with the active site are guided by the spatial arrangement and chiralities of the active site residues near the bound molecule. The favorable interaction with the amino acid residues with the enantiomer is responsible for the preferred binding of S-warfarin.

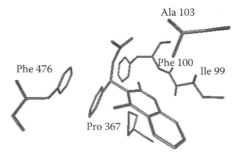

Figure 2.3 The residues that are in close proximity to the warfarin (1OG2. PDB) and interact with warfarin are shown. The image was generated using VMD (Humphrey, W. et al., 1996. *J. Mol. Graphics*, 14: 33).

It is interesting to note that the interchange of S-warfarin to R-warfarin in the binding site may not be significantly unfavorable. The stereochemical inversion of S-warfarin to R-warfarin by the interchange of the oxo group with the hydrogen atom attached to the chiral center in the binding site is expected to be associated with minimal loss of a favorable interaction. However, such an interchange may not be sterically unfavorable as there are rooms close to the binding site for such a stereochemical inversion. This suggests that the loss of hydrogen-bonding interaction with the oxo group of warfarin and Ala103 or unfavorable interaction with a second ligand bound in the available space (as suggested in the crystallographic study) could be responsible for the chiral discrimination. The former possibility is more likely and is consistent with the enzymatic stereopreference based on electrostatic interaction (Warshel et al., 2006, 3210).

Due to the large active site (approximately 470 Å³), it is possible that additional small molecules may simultaneously bind within the active site. It is also possible that some ligands that bind in the warfarin-binding pocket are prevented from moving closer to the heme and could behave as competitive inhibitors by occupying this binding site. There is space for a second drug molecule to bind at the heme when S-warfarin is bound. Molecular modeling suggests that either a second S-warfarin molecule or a different molecule, such as fluconazole, could simultaneously bind to the active site.

Simultaneous binding of more than a molecule is an interesting feature of the active site of CYP2C9. Warfarin and phenprocoumon (RS 4 hydroxy-3-(1-phenylpropyl)-2H-chromen-2-one) are structurally similar oral anticoagulants. Phenprocoumon is a substrate for CYP2C9. The chemical structure of phenprocoumon is shown in Figure 2.4. The observed differences in metabolic behavior and drug interaction of the two compounds are explored (He et al., 1999, 16).

Comparison of the metabolic behavior of S- and R-warfarin, S- and R- phenprocoumon, and different tautomeric forms of the structural mimics such as S- and R- pairs of enantiomers of 4-methoxyphenprocoumon,

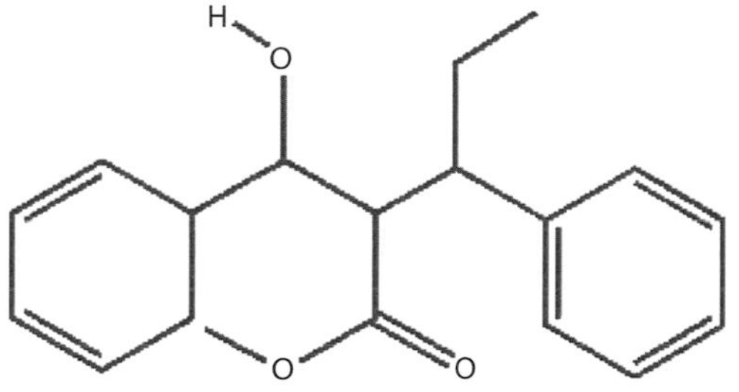

Figure 2.4 Chemical structure of phenprocoumone (RS)-4-hydroxy-3-(1-phenyl-propyl)-2H-chromen-2-one.

2-methoxyphenprocoumon, 4-methoxywarfarin, 2-methoxywarfarin, and 9-cyclocoumarol with CYP2C9 strongly suggests that the ring-closed form of S-warfarin and the ring-opened anionic form of S-phenprocoumon are the major and specific structural forms of the two drugs that interact with the active site residues of CYP2C9. The conclusion that S-warfarin and S-phenprocoumon interact with CYP2C9 in very different structural states is consistent with the significant differences in their metabolic profiles.

Suzuki and coworkers investigated the active sites of CYP2C19 and CYP2C9 by studying the inhibitor potencies of three series of *N*-3 alkyl-substituted phenytoin, nirvanol, and barbiturate derivatives. The interaction of the ligands with the active site revealed the interesting influence of chirality. All compounds were found to be competitive inhibitors of both enzymes, although the degree of inhibitory potency was generally much greater toward CYP2C19. In contrast, stereochemistry was an important factor in determining inhibitor potency toward CYP2C19 (Suzuki et al., 2004, 1). S-(+)-*N*-3-benzylnirvanol and R-(–)-*N*-3-benzylphenobarbital emerged as the most potent and selective CYP2C19 inhibitors. Both inhibitors were metabolized preferentially at their C-5 phenyl substituents. This may suggest that CYP2C19 prefers to orient the N-3 substituents away from the active oxygen species. Two residues that interact strongly with the bulky portion of the substrates in CYP2C19 are Val208 and Ile362. These residues are different in CYP2C9 (Leu208 and Leu362 in the latter case), but such subtle differences may not be responsible for the pronounced stereoselectivity observed for CYP2C19 inhibition. Potential electrostatic interactions that influence the stereoselective inhibitor binding to CYP2C19 are yet to be studied in detail. Metabolic profiling and homology modeling studies suggest that the two highest affinity inhibitors, S-*N*-3-benzylnirvanol and R-*N*-3-benzylphenobarbital, are preferentially bound in the active site of CYP2C19 with their C-5 phenyl groups oriented toward the active oxygen and *N*-3 benzyl groups bound within a lipophilic pocket made up of Ala, Val, Phe, Ile, Leu, and Phe. The inhibitor binding to CYP2C19, but not CYP2C9, was highly stereoselective.

Stereoselective oxidation of metoprolol with CYP2D6 was investigated by the heterologous expression of the corresponding complementary DNAs in the yeast *Saccharomyces cerevisiae* (Ellis et al., 1996, 647). The chemical structure of metoprolol is shown in Figure 2.5. O-demethylation is R-enantioselective. The α-hydroxylation showed a preference for S-metoprolol. The stereoselective properties of CYP2D6-Val are consistent with those observed for CYP2D6 in human liver microsomes. The difference in the stereoselective properties of CYP2D6-Val and CYP2D6-Met were analyzed with respect to a homology model of the active site of CYP2D6.

Optical spectroscopy is used to study the effects of mutations at the putative distal site of cytochrome P450 1A2 on chiral

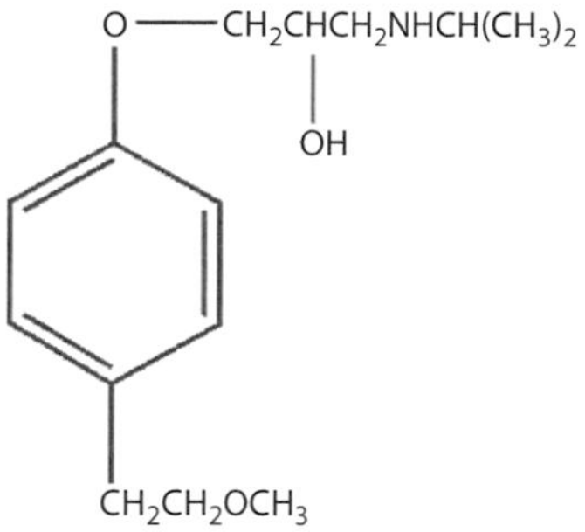

O —— CH₂CHCH₂NHCH(CH₃)₂

$$O\text{---}CH_2CHCH_2NHCH(CH_3)_2$$

OH

CH₂CH₂OCH₃

Figure 2.5 Chemical structure of metoprolol.

discrimination for binding R- and S-1-(1-naphthyl)-ethylamine, R- and S-1-cyclohexylethylamine, and R- and S-1-(4-pyridyl) ethanol (Krainev et al., 1993, 1951). The dissociation constants (K_d) are different for the R- and S-enantiomers of these ligands. The R/S ratios of the K_d values are 5.2 and 2.9 for 1-(1-naphthyl) ethylamine and 1-cyclohexylethylamine, respectively. The S/R ratio for 1-(4-pyridyl) ethanol is 6.0. Mutations at the putative distal site, such as Glu318Asp and Glu318Ala, remarkably enhanced the discrimination for 1-(1-naphthyl) ethylamine, which is reflected in the higher R/S ratio of the K_d values while the R/S ratio for 1-cyclohexylethylamine decreased. The R/S ratios are not changed for 1-(4-pyridyl) ethanol with Glu318Asp mutation and are markedly decreased by the Glu318Ala mutation. Thr319Ala mutation slightly increased the R/S ratio of the K_d values for 1-(1-naphthyl) ethylamine but markedly decreased the R/S ratio of both for 1-cyclohexylethylamine and 1-(4-pyridyl) ethanol. Differences between the R- and S-enantiomers of the standard enthalpy and entropy of 1-(4-pyridyl) ethanol binding are changed most remarkably by the Thr319Ser mutation. Although it is proposed that Glu318 and Thr319 play important roles in the chiral recognition and discrimination of dissymmetric ligands, it is difficult to conclusively predict the role of the specific residues due to the unavailability of the analysis of active site structure.

2.2 Enantioselectivity of hydride transfer of NADPH by alcohol oxidoreductase and conversion of epoxide to β-keto acid by 2-[(R)-2-hydroxypropylthio]-ethanesulfonate dehydrogenase

The alcohol oxidoreductase family contains a large number of enzymes showing diverse substrate specificities, all of which transfer a hydride between the 4-position of reduced nicotine adenine dinucleotide (NADPH)

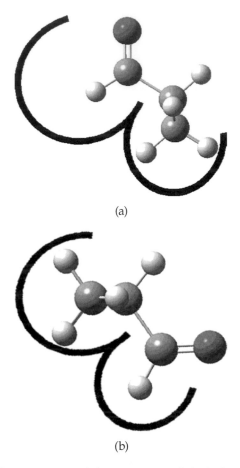

(a)

(b)

Figure 2.6 The schematic view of the active site of alcohol oxidoreductase with two lobes.

cofactor and a carbonyl function. Reduction of ethanal to ethanol by alcohol dehydrogenase is one of the oldest biotechnologies used.

The alcohol oxidoreductase (ADH) enzyme transfers the hydride between the 4-position of reduced nicotine adenine dinucleotide (NADPH) cofactor and the carbonyl functional group of substrate (aldehyde or ketone) (Kwiecien et al., 2009, 42). The introduction of the hydride ion from the re- or si-face of the substrate aldehyde results in an alcohol of either (pro-R) or (pro-S) configuration.

The specificity and selectivity of the enzyme depends on the geometry of the active site. The active site of the ADH is composed of two lobes: a big binding pocket and a small binding pocket. The schematic structure of the active site with two lobes is shown in Figure 2.6. The substrate can be positioned in two ways in the active site. In one orientation,

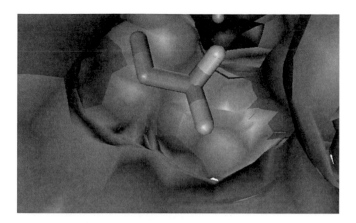

Figure 2.7 Active site of alcohol oxidoreductase of *S. cerevisiae* (PDB code: 2HCY) (Plapp et al. 2006, available from http://www.pdb.org/). The image was generated using VMD (Humphrey, W. et al.. 1996. *J. Mol. Graphics*, 14: 33).

the small pocket is occupied by the hydrogen atom and the big pocket is occupied by the alkyl group. In another orientation of the substrate, the hydrogen atom is in the big pocket and the alkyl group is forced to fit into the small pocket. The big pocket of the *L. brevis* is composed of hydrophobic side chains of Ala, Leu, Val, Leu, Met, and an aromatic ring of Tyr. The small binding pocket is composed of NADH and the hydrophobic side chain of Leu, Glu, Tyr, and Tyr. The big binding pocket of the *S. cerevisiae* is made up of Trp, Met, and Tyr, whereas the small binding pocket is composed of Trp and NADH. The active site of the ADH from *S. cerevisiae* contains Zn (catalytic), which is tetra-coordinated by the carbonyl oxygen atom of substrate, Nε atom of His, and two sulfur atoms of two Cys moieties. The active site of alcohol oxidoreductase of *S. cerevisiae* (PDB code: 2HCY) (Plapp et al., 2006) is shown in Figure 2.7. The presence of the Zn in the active site of ADH of *S. cerevisiae*, and amino acids required for its coordination constricts the volume of the small binding pocket. Consequently, the small pocket of the active site of ADH is smaller for the *S. cerevisiae* in comparison to that of the *L. brevis*. As a result, the placement of the alkyl group in the small pocket is highly unfavorable for the *S. cerevisiae*.

The ONIOM (QM/MM) calculations show that the binding of the methyl group of ethanal into the small pocket requires high energy cost (~22 k.Cal.mol^{-1}) (Kwiecien et al., 2009, 42). The large energy difference indicates the possibility of the development of unfavorable short-range steric repulsion. On the other hand, a minimal energy difference (~1 k.Cal.mol^{-1}) is observed in case of *L. brevis* when the methyl group is bound into the small pocket rather than the big pocket of the active site. This result is consistent with the observed low stereospecificity of the hydride transfer to the ethanal in *L. kefir*, which favors si-face attack by about a 3:1 ratio. The result indicates that the ease of hydride transfer to the re- or si- face of the substrate is dependent on the difference in the energies associated with the positioning of the substrate aldehyde in the two binding pockets. The propanal can adopt two different orientations in the active site of ADH of *L. brevis*, whereas the inverted orientation of the propanal in the active site of ADH of *S. cerevisiae* is impossible. The inverted orientation of propanal in the active site of ADH of *L. brevis* is associated with the ~6 k.Cal. mol^{-1} energy cost. This indicates that the re-face attack is strongly unfavorable for the substrate. The experimental study also shows that the ADH enzyme synthesizes a number of chiral alcohols in high enantiomeric excess (94–99%). This NADPH-requiring enzyme transfers the pro-R hydride from the cofactor to the si- face. Thus, computational studies confirm the experimental result and show that the stereospecificity and selectivity arise from the binding pattern of the substrate aldehyde in the active site of ADH.

Bacterial alcohol dehydrogenase 2-[(R)-2-hydroxypropylthio]-ethanesulfonate dehydrogenase (R-HPCDH) is an essential enzyme in the pathway of propylene metabolism in *Xanthobacter autotrophicus* strain Py2 (Clark et al., 2004, 6763). In this reaction, the reversible oxidation of 2-[(R)-2-hydroxypropylthio]-ethanesulfonate (R-HPC) to 2-(2-ketopropylthio) ethanesulfonate (2-KPC) is carried out, and is a major step in the bacterial conversion of chiral epoxides to β-keto acids. R-dehydrogenase is highly specific for the R-enantiomer of HPC and exhibits stringent chiral discrimination. Interestingly, a separate enzyme, S-HPCDH, catalyzes the oxidation of the corresponding S-enantiomer. Propylene is converted into a 95:5 enantiomeric mixture of R- and S-epoxypropane via an alkene monooxygenase. Enantiomers of epoxypropane are subsequently ring-opened by the thiol of the atypical cofactor coenzyme M (CoM) in a reaction catalyzed by a coenzyme M transferase (a zinc-dependent epoxyalkane). The resulting products of this reaction are R- and S-HPC.

It is proposed that a binding pocket for the alkylsulfonate moiety of the substrate plays a decisive role in the stereospecificity exhibited by R-HPCDH. The capacity to properly orient the CoM functional group is important. It is proposed that either enantiomer will bind with the same high affinity but that only the R-enantiomer is oriented properly for catalysis. Kinetic and biochemical studies strongly suggest that the sulfonate of CoM is integral to the binding of substrate and influences the enantioselectivity of R-HPCDH. It is speculated that the Arg residues (such as Arg152 and Arg196) of R-HPCDH are also crucial to the recognition and binding of the sulfonate of CoM. The electrostatic interactions between the positively charged arginine side chains and the negatively charged oxygen atoms of the substrate sulfonate group could have an important role in the stereocontrol. The negatively charged sulfonate group can locate and orient the substrate molecule for binding to the active site. The proposed salt bridges formed by these interactions may also orient residues within the active site for optimal catalysis and induce conformational changes in the active site.

The basis of chiral discrimination by the hydroxypropylthio-ethanesulfonate dehydrogenase is analyzed from crystal structure analysis (Krishnakumar et al., 2006, 8831). As concluded from the earlier study (Clark et al., 2004, 6763), substrate sulfonate binding is the key to orienting the substrate at the active site for hydride abstraction. The structures of both R-HPCDH and 2-ketopropyl-CoM oxidoreductase/carboxylase (2-KPCC) illustrate the effectiveness of CoM as a cofactor in this biochemical pathway. Sulfonate binding being the key to correct substrate alignment, the sulfonate binding and the specific sulfonate binding sites are expected to be different in R-HPCDH and S-HPCDH. However, direct crystallographic analysis of S-HPCDH has not been carried out, but a homology model has been constructed for S-HPCDH. The Arg residues (Arg152 and Arg196) that bind the sulfonate in R-HPCDH are replaced in S-HPCDH with Met147 and Gln191, respectively. The latter side chains are not capable of binding the substrate as occurs in R HPCDH. Effective hydride abstraction by short-chain dehydrogenases is dependent on the correct orientation of the hydride with respect to the nicotinamide and the correct orientation of the substrate hydroxyl group with respect to the Tyr of the catalytic triad. To achieve the proper position and the orientation of the substrate hydroxyl group, the spatial orientation of the sulfonate and methyl groups needed to be different in the two enzymes. The orientations of active site residues and the NAD⁺ are the same in the two active sites. The sulfonate and the methyl groups are switched to position the hydrogen and hydroxyl groups of the substrate for hydride and proton abstraction, respectively. The sulfonate binding sites in R- and S-HPCDH are analogous. However, they are composed of different sets of amino acid residues. These sets of residues are located in spatially different positions in the respective enzymes. It has been shown that the orientations of active site residues and the NAD⁺ are the same, but the sulfonate and the methyl groups are switched to position the hydrogen and hydroxyl groups of the substrate for hydride and proton abstraction, respectively. The differences in the positively charged residues within the substrate binding region of the enzyme could assist in controlling the chiral discrimination.

2.3 Lipoxygenase and cyclooxygenase: generation of chiral peroxide from achiral polyunsaturated fatty acid

Lipoxygenases (abbreviated as LOX) are a family of iron-containing enzymes found in plants, animals, and fungi that catalyze the dioxygenation of achiral polyunsaturated fatty acids in lipids, leading to the formation of fatty acid hydroperoxide. Both LOX and cyclooxygenase activity involve free radical chemistry. Products of lipoxygenases are involved in diverse cell functions. The iron atom in lipoxygenases is bound by four ligands, three of which are histidine residues. Six histidines are conserved in all lipoxygenase sequences. The active site iron is coordinated by εN of three conserved His residues and one oxygen of the C-terminal carboxyl group. Cyclooxygenase (abbreviated as COX) is an enzyme that is responsible for the formation of important biological mediators called prostanoids, including prostaglandins, prostacyclin, and thromboxane. Pharmacological inhibition of COX can provide relief from the symptoms of inflammation and pain. Nonsteroidal anti-inflammatory drugs, such as aspirin and ibuprofen, exert their effects through inhibition of COX.

The active site is composed of two helices that cross at the active site and include internal stretches of π-helix that provide three histidine (His) ligands to the active site iron. Two cavities in the major domain of soybean lipoxygenase-1 extend from the surface to the active site. The funnel-shaped cavity is suggested to function as a dioxygen channel. The long narrow cavity is suggested to be a substrate pocket. Mammalian enzyme is more compact and contains only one boot-shaped cavity. A third cavity runs from the iron site to the interface of the β-barrel and catalytic domains in soybean lipoxygenase-3. These cavities and the iron site form a continuous passage within the protein structure.

LOX and COX react with an achiral polyunsaturated fatty acid and oxygen to form a chiral peroxide product of high regio- and stereochemical purity (Schneider et al., 2007, 473). Exactly how both dioxygenases achieve this positional and stereo control to form the respective product is yet to become clear. Four mechanistic models are presented to account for the specific reactions of molecular oxygen with a fatty acid in the LOX or COX active site. The LOX and COX enzymes employ different strategies to activate the fatty acid substrate and thus facilitate reaction with O_2. The proteins themselves require oxidation from the resting state to the catalytically active form. In both cases, enzyme activation is accomplished by trace amounts of fatty acid hydroperoxides. A Fe^{2+} to Fe^{3+} oxidation generates a ferric iron- Fe^{3+}-hydroxy species in LOX enzymes. In COX enzymes, oxidation at the peroxidase active site is conducted through the protein to generate a tyrosyl radical (Tyr385) in the oxygenase active site.

The subsequent reactions on their fatty acid substrate share some similarity. Stereoselective hydrogen abstraction from the substrate occurs in both enzymes, which are oxidized to a free radical capable of reaction with unactivated molecular oxygen. The hydrogen is removed from the methylene carbon of a selected 1,4-cis,cis pentadiene unit on the fatty acid substrate. The hydrogen abstraction is followed by regioselective oxygen addition on the opposite face of the substrate to form a fatty acid peroxyl radical as the primary oxygenation product.

Different possible mechanisms of oxygenation control of arachidonic acid that may be employed by LOX and COX enzymes have been suggested. It has also been indicated that two or more mechanisms may act in concert. It is possible that the fatty acid substrate is bound in such a way that only one of the reactive carbon centers of the pentadienyl radical is accessible to molecular oxygen due to steric shielding. Oxygen might have free access to some parts of the active site, but is expected to be denied access to other regions by the protein structure. The concept of an oxygen pocket is closely related to steric shielding. It is possible that there could be a pocket within the active site where oxygen can reside due to favorable solubility properties and available space such that it can react with the fatty acid radical intermediate in the intended positional and stereochemical configuration. The steric shielding is involved in controlling oxygenation and could be the exclusive mechanism. It has also been suggested that the protein may direct molecular oxygen through a channel to the desired site of oxygenation on the activated fatty acid substrate. This ensures the positional specificity and stereospecificity.

An important question is how far the fatty acid enters the LOX active site. This is related to the position of the cis-cis double bonds on the fatty acid chain that are targeted for reaction, and the head-to-tail orientation of the fatty acid in the active site is a further determinant of reaction specificity. It is proposed that oxygenations with S-stereochemistry occur deep in the active site, relative to the point of substrate entry and the catalytic iron. On the contrary, all oxygenations with R stereochemistry occur on the proximal or near side of the iron. The Ala-to-Gly mutation in different S-LOX and Gly to Ala in R-LOX are found to switch the position and stereochemistry of oxygenation across one face of the reacting pentadiene significantly. It is difficult to understand how removal or addition of a methyl group can switch specific oxygenation from one end of the reacting pentadiene to the other. The Ala-to-Gly mutation is expected to make more space available and not to block a channel. The additional space near the Gly is associated with R-oxygenation nearby. It is far from clear that how this factor unfavors the reaction at the deep end of the active site where the normal S-oxygenation occurs. EPR studies point to a change in the ligand environment of the catalytic iron, presumably induced by conformational adjustments within the protein.

Through molecular dynamics modeling of the oxygen movement, however, a different channel for oxygen was identified, and this appeared to deliver O_2 to the appropriate place for reaction in the 15S position of arachidonic acid. The x-ray structures of COX containing bound fatty acid substrate have helped direct mutagenesis studies and served as a platform for molecular dynamics simulations. In a COX enzyme containing a catalytically inert cobalt heme group, arachidonic acid is bound in the oxygenase active site with the critical pro-S hydrogen at C-13 suitably positioned for hydrogen abstraction by an incipient Tyr385 radical. During catalysis, this hydrogen removal is followed by 11R oxygenation, and all the experimental evidence from product analyses suggests that the stereocontrol of this step is stringent. Molecular dynamics simulations support steric shielding being important for 11R oxygenation. The 11S face of the pentadienyl radical (the hydrogen abstraction side) is well shielded from access of molecular oxygen. The distance between carbons C_{11}, C_{13}, and C_{15} of a planar arachidonate-derived pentadienyl radical and the nearest protein atom is less than 1 Å on an average. The simulations indicate an open space on the pro-R side of C_{11}, providing sufficient room for the access of oxygen. It is suggested that oxygen resides in an oxygen pocket on the pro-R side of C_{11}, and MD calculations provide support for the existence of the putative oxygen pocket.

The steric shielding hypothesis may not explain the overwhelming large switch in stereocontrol induced by a simple Ser530Thr substitution. The conformation of the reactant could be a determinant of 15R/15S specificity as noted from two mutant forms of COX-2 (Leu384Phe and Gly526Ser). It is proposed that the structure of the active site is not designed for synthesis of a particular stereochemistry. It is possible that the formation of 15R or 15S-hydroperoxy products may depend on the induced conformation of the omega side chain of the reacting intermediate. The major difficulty in understanding how the LOX and COX enzymes control the regio- and stereochemistry of their catalytic reactions is the mechanism of access of the second substrate, oxygen, to the reactive intermediate.

2.4 Nitric oxide synthase: effects of substrate and cofactors on chiral discrimination for binding the enantiomeric ligands

Nitric oxide synthases (EC 1.14.13.39) (NOSs) are a family of eukaryotic enzymes that catalyze the production of nitric oxide (NO) from L-arginine. NO is an important cellular signaling molecule, having a vital role in many biological processes. Nakano and coworkers studied the effects of substrate, L-Arg and cofactors, 6R-l-erythro-5,6,7,8-tetrahydrobiopterin and calmodulin, on chiral discrimination by rat neuronal nitric oxide synthase (nNOS) for binding the enantiomers of different ligands. The ligands considered are 1-(1-naphthyl)ethylamine (ligand I), 1-cyclohexylethylamine (ligand II), and 1-(4-pyridyl)ethanol (ligand III) under anaerobic conditions by optical absorption spectroscopy (Nakano et al., 1998, 767). The ratio of the dissociation constant (K_d) values for the S- and R-enantiomers of ligand I (S/R) was 30, while the S/R ratio for ligand II and the R/S ratio for ligand III were 1.8 and < 0.14, respectively. However, in the presence of 1 mM L-Arg, the S/R ratio of the K_d values for ligand I was decreased down to 5.9. In the presence of both 1 mM L-Arg and 0.1 mM H4B, the S/R ratios for ligands I and II and the R/S ratio for ligand III were enormously increased up to 29, more than 80, and 60, respectively. These and other spectral observations strongly suggest that strict chiral recognition at the active site of nNOS during catalysis is exhibited only in the presence of the active effector. However, the influence of active site residues on discrimination is yet to be understood.

2.5 Enoyl reductase: chirality dependent branching of a growing polyketide chain

Kwan and coworkers (Kwan et al., 2008, 1231) observed that when an enoyl reductase (ER) enzyme of a modular polyketide synthase (PKS) reduces a propionate extender unit that has been newly added to the growing polyketide chain, the resulting methyl branch might have either S or R configuration. A correlation is noted between the presence or absence of a unique tyrosine residue in the ER active site and the chirality of the methyl branch that is introduced. When this position in the active site is occupied by a tyrosine residue, the methyl branch has an S-configuration, otherwise it has an R-configuration. A mutation (Tyr to Val) in an erythromycin PKS-derived ER in a model PKS in vivo caused a switch in the methyl branch configuration in the product from S to R. In contrast, alteration (Val to Tyr) at this position in a rapamycin-derived PKS ER was insufficient to achieve a switch from R to S, showing that additional residues also participate in stereocontrol of enoylreduction.

References

Atkins, W.M. and Sligar, S.G. (1988). The roles of active site hydrogen bonding in cytochrome P-450cam as revealed by site-directed mutagenesis. *J. Biol. Chem.*, 263: 18842–18849.

Atkins, W.M. and Sligar, S.G. (1989). Molecular recognition in cytochrome P-450: Alteration of regioselective alkane hydroxylation via protein engineering. *J. Am. Chem. Soc.*, 111: 2715–2717.

Clark, D.D., Boyd, J.M., and Ensign, S.A. (2004). The stereoselectivity and catalytic properties of *Xanthobacter autotrophicus* 2-[(R)-2-Hydroxypropylthio]ethane-sulfonate dehydrogenase are controlled by interactions between C-Terminal Arginine residues and the sulfonate of coenzyme M. *Biochemistry*, 43: 6763–6771.

Ellis, S.W., Rowland, K., Ackland, M.J., Rekka, E., Simula, A.P., Lennard, M.S., Wolf, C.R., and Tucker, G.T. (1996). Influence of amino acid residue 374 of cytochrome P-450 2D6 (CYP2D6) on the regio- and enantio-selective metabolism of metoprolol. *Biochem. J.*, 316: 647–654.

Flockhart, D.A. (2007). Drug interactions: Cytochrome P450 drug interaction table. Indiana University School of Medicine. <http://medicine.iupui.edu/clin-pharm/ddis/table.asp> (accessed June 19, 2010).

Furuya, H., Shimizu,T., Hirano, K., Hatano, M., Fujii-Kuriyama, Y., Raag, R., and Poulos, T.L. (1989). Site-directed mutageneses of rat liver cytochrome P-450$_d$: Catalytic activities toward benzphetamine and 7-ethoxycoumarin. *Biochemistry*, 28: 6848–6857.

Gerber, N.C. and Sligar, S.G. (1992). Catalytic mechanism of cytochrome P-450: Evidence for a distal charge relay. *J. Am. Chem. Soc.*, 114: 8742–8743.

Gotoh, O. (1992). Substrate recognition sites in cytochrome P450 family 2 (CYPB) proteins inferred from comparative analyses of amino acid and coding nucle-otide sequences. *J. Biol. Chem.*, 267: 83–90.

Graham-Lorence, S., Khalil, M.W., Lorence, M. C., Mendelson, C.R., and Simpson, E.R. (1991). Structure-function relationships of human aromatase cyto-chrome P-450 using molecular modeling and site-directed mutagenesis. *J. Biol. Chem.*, 266: 11939–11946.

Hanioka, N., Gonzalez, F.J., Lindberg, N.A., Liu, G., Gelboin, H.V., and Korzekwa, K.R. (1992). Site-directed mutagenesis of cytochrome P450s CYP2A1 and CYP2A2: Influence of the distal helix on the kinetics of testosterone hydroxy-lation. *Biochemistry*, 31: 3364–3370.

He, M., Korzekwa, K.R., Jones, J.P., Rettie, A.E., and Trager, W.F. (1999). Structural forms of phenprocoumon and warfarin that are metabolized at the active site of CYP2C91. *Arch. Biochem. Biophys.*, 372: 16–28.

Humphrey, W., Dalke, A., and Schulten, K. (1996). VMD: Visual molecular dynam-ics. *J. Mol. Graphics*, 14: 33–38.

Imai, M., Shimada, H., Watanabe, Y., Matsushima-Hibiya, Y., Makino, R., Koga, H., Horiuchi, T., and Ishimura, Y. (1989). Uncoupling of the cytochrome P-450cam monooxygenase reaction by a single mutation, threonine-252 to alanine or valine: A possible role of the hydroxy amino acid in oxygen activa-tion. *Proc. Natl. Acad. Sci. (USA)*, 86: 7823–7827.

Imai, Y. and Nakamura, M. (1989). Point mutations at threonine-301 modify substrate specificity of rabbit liver microsomal cytochromes P-450 (laurate (ω-1)-hydroxylase and testosterone 16α-hydroxylase). *Biochem. Biophys. Res. Commun.*, 158: 717–722.

Ishigooka, M., Shimizu, T., Hiroya, K., and Hatano, M. (1992). Role of Glu318 at the putative distal site in the catalytic function of cytochrome P450$_d$. *Biochemistry*, 31: 1528–1531.

Iwasaki, M., Juvonen, R., Lindberg, R., and Negishi, M. (1991). Alteration of high and low spin equilibrium by a single mutation of amino acid 209 in mouse cytochromes P450. *J. Biol. Chem.*, 266: 3380–3382.

Krainev, A.G., Shimizu, T., Ishigooka, M., Hiroya, K., and Hatano, M. (1993). Chiral recognition at cytochrome P450 1A2 active site: Effects of mutations at the putative distal site on the bindings of asymmetrical axial ligands. *Biochemistry*, 32: 1951–1957.

Krishnakumar, A.M., Nocek, B.P., Clark, D.D., Ensign, S.A., and Peters, J.W. (2006). Structural basis for stereoselectivity in the (R)- and (S)-hydroxypropylthioethanesulfonate dehydrogenases. *Biochemistry*, 45: 8831–8840.

Kwan, D.H., Sun, Y., Schulz, F., Hong, H., Popovic, B., Joalice, C., Sim-Stark, C. Haydock, S.F., and Leadlay, P.F. (2008). Prediction and manipulation of the stereochemistry of enoylreduction in modular polyketide synthases. *Chem. Biol.*, 15: 1231–1240.

Kwiecien, R.A., Ayadi, F., Nemmaoui, Y., Silvestre, V., Zhang, B.L., and Robins, R.J. (2009). Probing stereoselectivity and pro-chirality of hydride transfer during short-chain alcohol dehydrogenase activity: A combined quantitative 2H NMR and computational approach. *Arch. Biochem. Biophys.*, 482: 42–51.

Martinis, S.A., Atkins, W.M., Stayton, P.S., and Sligar, S.G. (1989). A conserved residue of cytochrome P-450 is involved in heme-oxygen stability and activation. *J. Am. Chem. Soc.*, 111: 9252–9253.

Meng, X.-Yu, Zheng, Q.C., and Zhang, H.-X. (2009). A comparative analysis of binding sites between mouse CYP2C38 and CYP2C39 based on homology modeling, molecular dynamics simulation and docking studies. *Biochim. Biophys. Acta*, 1794: 1066–1072.

Meunier, B., de Visser, S.P., and Shaik, S. (2004). Mechanism of oxidation reactions catalyzed by cytochrome p450 enzymes. *Chem. Rev.*, 104: 3947–3980.

Nakano, K., Sagami, I., Daff, S., and Shimizu, T. (1998). Chiral recognition at the heme active site of nitric oxide synthase is markedly enhanced by L-arginine and 5,6,7,8-tetrahydrobiopterin. *Biochem. Biophys. Res. Commun.*, 248: 767–772.

Plapp, B.V., Savarimuthu, B.R., and Ramaswamy, S. (2006). Available from http://www.pdb.org/.

Poulos, T.L., Finzel, B.C., and Howard, A.J. (1987). High-resolution crystal structure of cytochrome P450cam. *J. Mol. Biol.*, 195: 687–700.

Poulos, T.L., Finzel, B.C., Gunsalus, I.C., Wagner, G.C., and Kraut, J. (1985). The 2.6Å crystal structure of *Pseudomonas putida* cytochrome P-450. *J. Biol. Chem.*, 260: 16122–16130.

Raag, R., Martinis, S.A., Sligar, S.G., and Poulos, T.L. (1991). Crystal structure of the cytochrome P-450cam active site mutant Thr252Ala. *Biochemistry*, 30: 11420–11429.

Schneider, C., Pratt, D.A., Porter, N.A., and Brash, A.R. (2007). Control of oxygenation in lipoxygenase and cyclooxygenase catalysis. *Chem. Biol.*, 14: 473–488.

Suzuki, H., Kneller, M.B., Rock, D.A. Jones, J.P., Trager, W.F., and Rettiea, A.E. (2004). Active-site characteristics of CYP2C19 and CYP2C9 probed with hydantoin and barbiturate inhibitors. *Arch. Biochem. Biophys.*, 429: 1–15.

Warshel, A., Sharma, P.K., Kato, M., Xiang, Y., Liu, H., and Olsson, M.H.M. (2006). Electrostatic basis for enzyme catalysis. *Chem. Rev.*, 106: 3210–3235.

Williams, P.A., Cosme, J., Ward, A., Angove, H.C., Vinkovic, D.M., and Jhoti, H. (2003). Crystal structure of human cytochrome P450 2C9 with bound warfarin. *Nature*, 424: 464–468.

Zhou, D., Korzekwa, K.R., Poulos, T., and Chen, S. (1992). Site-directed mutagenesis study of human placental aromatase. *J Biol. Chem.*, 267: 762–768.

chapter three

Transferases and chiral discrimination

Transferases are enzymes that transfer a group from a donor compound to another acceptor compound. The groups commonly transferred by various transferases are alkyl or aryl groups, single carbon groups, aldehyde or ketonic groups, acyl groups, nitrogenous groups, phosphorus-containing groups, sulfur-containing groups, and selenium-containing groups. As per the recommended names, these enzymes could be termed either *acceptor-group transferase* or *donor-group transferase*. If the chemical structure of the group being transferred is chiral, a number of transferases efficiently exclude the possibility of the incorporation of a particular chiral form (in many cases, the corresponding nonnatural form) of the group (when chirality is present in the respective group). For example, the active site of the peptidyl transferase at ribosome specifically incorporates the L-amino acid and it is practically impossible to misincorporate D-enantiomer. Such selectivity is important for the fidelity of the biochemical reaction so that the product with natural chirality is generated rather than the incorporation of the wrong enantiomer. This is essential for the production of the fully functional protein structure. Although the availability of the D-enantiomer is negligible in today's predominantly homochiral world with the exclusive presence of L-amino acid in many organisms, it is still important to arrive at an understanding of the origin of the fidelity mechanism. The active site structure of the organisms is the structure that developed chiral selectivity through evolution. Understanding of the fidelity mechanism is expected to help design strategies for targeted protein synthesis. In this chapter, we discuss the features of chiral discrimination exhibited by peptidyl transferase, telomerase, HIV-1 reverse transcriptase, and nuclear DNA polymerase.

3.1 Peptidyl transferase center within ribosome: peptide bond formation and chiral discrimination

Protein synthesis is a multistep reaction and leads to the polymerization of amino acids to finally generate the protein in its functional form (Berg et

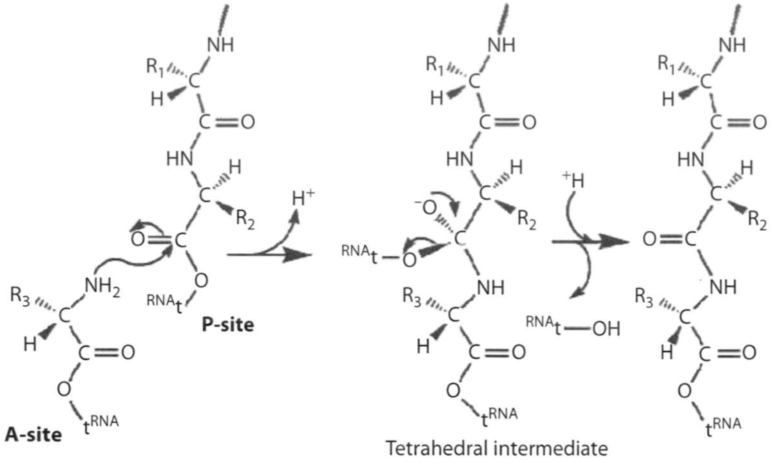

Figure 3.1 Scheme of peptide bond formation at the peptidyl transferase center in ribosome (Thirumoorthy, K. and Nandi, N. 2006. *J. Phys. Chem. B.*, 110: 8840. Reprinted with permission from the American Chemical Society. © American Chemical Society.)

al., 2003, 814). Out of the several partial reactions of protein biosynthesis, the peptidyl transferase reaction is one of the most intensively studied ones. Peptidyl transferase is an aminoacyl transferase enzyme (EC 2.3.2.12) in ribosome and is essentially a ribozyme. Peptide bond formation occurs in the ribosome, which is a large and complex macromolecule (approx 2.5 MDa in bacteria) and consists of two subunits (small subunit, 30S and large subunit, 50S). The active site is termed the peptidyl transferase center (PTC). Out of the five steps in protein synthesis, the PTC performs the elongation of the protein chain after the activation of amino acid and after the formation of the initiation complex. A scheme of the peptide bond formation involving A- and P-terminals occurring within PTC is shown in Figure 3.1. In this process, the nucleophilic α-amino group of the aminoacyl-tRNA at A-site makes a nucleophilic attack at the electrophilic carbonyl carbon of the ester bond of the peptidyl-tRNA at the P-site and forms an intermediate, which subsequently generates the peptide bond, and the polypeptide chain length grows (Berg et al., 2003, 814). Although ribosome is made up of RNA and proteins, the peptidyl transferase center is essentially composed of nucleic acid with no protein present within 1.5 nm of the active site. The active site of the large subunit of the ribosome is composed of layers of conserved nucleotides (Youngman et al., 2004, 589). This is in contrast to the more common occurrence of amino acid residues rather than nucleotides in the active sites of various enzymes. Schematic representations of the active site of architecture of the peptidyl transferase center taken from the crystal structure of the ribosomal part of *Haloarcula*

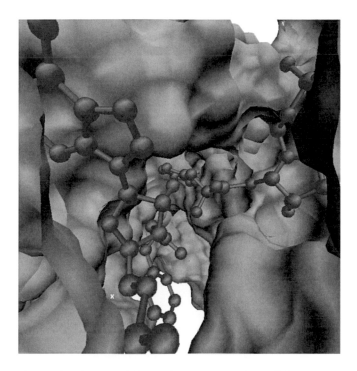

Figure 3.2 **See color insert.** Schematic representation of the active site of the architecture of peptidyl transferase center (PTC) taken from the crystal structure of the ribosomal part of *Haloarcula marismortui* (Hansen, J.L. et al., 2002. *Proc. Natl. Acad. Sci. (USA).* 99: 11670). The crystal structure of CCA-Phe-cap-biotin bound simultaneously at half-occupancy to both the A-site and P-site of the 50S ribosomal subunit (1Q86.PDB) is used to generate the structure. The reactants (amino (A) and peptidyl (P) terminals) are shown by ball and stick representation, and the enclosing nanospace (cavity of the active site) is shown as a gray surface.

marismortui are shown in Figure 3.2 (different representations of the same structure of the PTC are shown in the introduction section in Figures 1.8a and 1.8b, respectively).

How chiral discrimination is carried out by PTC and how the ribosomal pathway achieves the exclusion of amino acids other than the L-enantiomer are important questions. Although the racemic compound (crystalline) is more stable than the enantiomeric form, racemization does not occur in the natural biosynthetic pathway. The mechanism of retention of enantiopurity is not straightforward to understand since D-amino acid is involved in nonribosomal peptide synthesis, posttranslational modifications, and is present in bacterial cell walls and the human brain, for example. Despite the reaction being remarkably fast, the chance of incorporation of wrong (D-enantiomer, for example) amino acid is only 10^{-4} per amino acid for a 1000 residue protein in the ribosomal pathway. Modification of active site structure (mutation) is necessary to enhance D-amino acid incorporation during elongation. Several experimental and computational studies suggest that accuracy in chiral discrimination in ribosomal peptide synthesis is controlled by the chiral amino acid terminals, chiral sugar rings involved, and structure of the PTC, including the achiral bases as discussed later.

A number of experimental studies focus on discrimination in peptide synthesis. However, various experimental results on chiral discrimination in peptide synthesis must be interpreted with caution. The states of the reacting amino acid segments (their degrees of freedom, proximity, the mutual orientation of the amino and peptidyl terminal, and interaction with the surrounding residues) might differ from one experimental situation to the other. Different relative spatial arrangements of the reacting moieties and the proximity to neighboring residues as well as varying degrees of confinement during the reaction may give rise to variations in intermolecular interaction and the resulting discriminations.

For example, various results concerning the peptide bond formation reaction are obtained using different analogues in the in vitro experiments. The substrate specificity of the peptide synthesis process, especially as regards the structural nature and spatial orientation of the aminoacyl moiety, has been investigated using analogues of puromycin, aminoacylated nucleotides structurally related to the 3' terminus of *t*RNA (Monro and Marcker, 1967, 347; Monro et al., 1968, 1042; Bhuta et al., 1981, 8; Quiggle et al., 1981, 3480), and tRNA analogues in which the cognate amino acid was maintained in a "defined" spatial orientation (Hecht, 1977, 1671; Sprinzl and Cramer, 1979, 1; Harris and Pestka, 1977, 413). The reason for using the mimics is that the incorporation of D-amino acid by aminoacylation (which is a prerequisite for the formation of aminoacyl-*t*RNA) is difficult or even impossible in many cases. As the D-amino acid could not be incorporated as a reacting terminal in the tRNA or its analogue, the synthesis of D-L peptide could not be performed in such cases. Hence, due to the foregoing reasons, the experimental results concerning chiral discrimination in peptide synthesis must be judged with caution.

In an early study, Calender and Berg observed that D-tyrosine could be esterified to *t*RNA[Tyr] and then could be incorporated into peptides (Calendar and Berg, 1967, 39). D-tyrosine was incorporated into the peptide in an assay for the protein synthesizing system. L-tyrosine incorporation is 114 μmoles, and D-tyrosine incorporation is only 19 μmoles. The reduced (about 1/6th of the L-isomer) incorporation of D-tyrosine indicates that protein synthetic system is unfavorable toward the incorporation of D-amino acids into peptides. Yamane and coworkers studied stereoselectivity in peptide chain elongation (Yamane et al., 1981, 7059). The overall rate of dipeptide formation is 30-fold higher for the L-enantiomer compared to the D-enantiomer. It is noted that 75% to 90% of D-Tyr-tRNA dissociates from the ribosome, while less than 30% of L-Tyr-tRNA is lost through this pathway. It is estimated that a 6-fold rate difference arises in binding of tRNA to the A site and a 5-fold rate difference arises in the peptide bond formation. These early experiments draw attention to the homochiral preference in peptide bond formation.

In a series of detailed studies, Hecht and coworkers studied the extent of incorporation of D-amino acid into ribosome (Heckler et al., 1983, 4492; Roesser et al., 1989, 5185; Heckler et al., 1988, 7254). Misacylated *t*RNAs such as *N*-acetyl-D-phenylalanyl-*t*RNA^{Phe}, *N*-acetyl-L-tyrosyl-*t*RNA^{Phe}, *N*-acetyl-D-tyrosyl-*t*RNA^{Phe}, and *N*-acetyl-β-phenylalanyl-*t*RNA^{Phe} are used to study the peptidyl (P-site) binding and peptide bond formation in a cell-free system employing *E. coli* ribosome programmed with poly-uri-dylic acid (Heckler et al., 1983, 4492; Roesser et al., 1989, 5185). Unlike the *t*RNA^{Phe} activated with L-phenylalanine and L-tyrosine, the *N*-acetyl-D-phenylalanyl-*t*RNA^{Phe} and *N*-acetyl-D-tyrosyl-*t*RNA^{Phe} performed poorly for dipeptide formation when L-phenylalanyl-*t*RNA^{Phe} acts as acceptor (A-site) *t*RNA. It, however, could not be established that this inability is due to the poor binding of these *t*RNAs with the ribosomal P-site or due to their inefficiency in participating in the peptide synthesis reaction. Another study confirms the foregoing results (Heckler et al., 1988, 7254). The relative yield of dipeptide synthesis for D-analogue is only 7% and 13% for *N*-acetyl-D-tyrosyl–tRNA^{Phe} relative to the L-*N*-acetyl phenyl-alanyl–tRNA^{Phe}. The small amount of dipeptide produced indicates that D isomer incorporation is severely inhibited compared to the natural iso-mer (Heckler et al., 1983, 4492). The further influence of the chiral structure of the amino acid moiety is noted for *N*-acetyl-(D, L) β-phenylalanyl–*t*RNA^{Phe} system. The relative yield of dipeptide synthesis for the analogue is about 113% relative to the L-*N*-acetyl phenylalanyl-tRNA^{Phe} system. The authors indicated that the normal position of the *N*-acetyl amino group on C^{α} may not be optimal for peptide bond formation, and modification of the chiral structure can enhance the efficacy of the reaction. The studies by Hecht and coworkers confirm the earlier studies of Calendar and Berg as well as that of Yamane and Hopfield that the incorporation of D-amino acids into a peptide is unfavorable relative to the incorporation of natural enantiomer. In order to understand the structural requirements for the participation as an acceptor in the reaction, Hecht and coworkers have used aminoacyl-tRNA analogues bearing noncognate aminoacyl moi-eties (Roesser et al., 1989, 5185). Two tRNA^{Phe}'s bearing noncognate amino acids (*N*-pyroglutamyl-L-*O*-methyltyrosyl-tRNA^{Phe} and *N*-pyroglutamyl-L-phenylglycyl-tRNA^{Phe}) are also able to participate as acceptors in the peptidyltransferase reaction. In contrast, neither *N*-pyroglutamyl-D-phenylalanyl-tRNA^{Phe} nor *N*-pyroglutamyl-D-tyrosyl-tRNA^{Phe} acted as an acceptor in the peptidyltransferase reaction following treatment with pyroglutamate aminopeptidase *N*-pyroglutamyl-D, L-β-phenylalanyl-tRNA^{Phe} produced dipeptide only to the extent of 8%.

Chamberlin and coworkers also noted the diminished activity of D-isomer (Bain et al., 1991, 5411). In an alternative approach, in the presence of nonsense codons incorporated in mRNA (codons that effect termination of peptide synthesis as there is no corresponding tRNAs) such as 5'UAG, 5'UAA, and 5'UGA, a compensatory DNA mutation can lead to the production of mutated tRNA (called suppressor tRNA) after activation with a normal amino acid. The presence of the suppressor tRNA during in vitro cell-free translation of mRNA containing a nonsense suppression site can incorporate nonnatural amino acids. While tRNAs activated with iodotyrosine, *N*-methylphenylalanine, or glycine functioned well in suppression but D-pheylalanyl-tRNA did not. The result that D-amino acid fails to get incorporated is further confirmed by both the normal rapid assay method as well as by the direct analysis of the isolated HPLC fraction.

A significant influence of the chirality of the sugar ring involved in the CCA moiety is also noted (Hecht, 1992, 545). This result is important in understanding the heterochiral relationship of D-sugar and L-amino acid in peptide synthesis, where the former is present as a part of the tRNA structure. While the *N*-acetyl-D-phenylalanyl-tRNA^{Phe} and *N*-acetyl-D-tyrosyl-tRNA^{Phe} produced only a small amount of dipeptide, the 2' and 3' deoxyadenosine analogues (L-*N*-acetyl phenylalanyl-tRNA^{Phe}–CC2'dA and L-*N*-acetyl phenylalanyl-tRNA^{Phe}–CC3'dA) produced no detectable dipeptide formation. The result indicates that the effect of the sugar ring chirality is decisive over the peptide bond formation and alteration of the chirality of the same is unfavorable for the synthesis. Suppression efficiencies of misacylated D-phenylalanyl-tRNA$^{Gly}_{CUA}$–dCA are nil as compared to other misacylated phenylalanyl-tRNA$^{Gly}_{CUA}$–dCA.

Chiral discrimination is also studied in different model systems. Puromycine [9-{3'-deoxy-3'-[(4-methoxy-phenylalanyl)amino]-β-D-ribofuranosyl}-6(*N*,*N*'-dimethylamino) purine] is a small molecule mimic of aminoacyl-tRNA (aa-tRNA) and acts as a translation inhibitor by entering the ribosomal A-site and participating in peptide bond formation with the nascent peptidyl chain. Puromycin blocks protein synthesis by acting as an analogue of the charged tRNA. It substitutes for an incoming aminoacyl-tRNA as the acceptor of the carboxyl-activated peptide and forms peptidyl-puromycin. While peptidyl-tRNA in ribosome is normally transferred to the amino group of the next aminoacyl-tRNA, the carboxyl-activated peptide is transferred to puromycin. This causes termination of the sequential extension of peptide and stops the growth of the nascent peptide chain (Nathans, 1964, 585; Nathans and Neidle, 1963, 1076). Note that the orientational degrees of freedom of A- and P- terminals in such analogues are in principle different from those within the ribosomal active site. Hence, the extent of discrimination might be different in these two cases.

While it is virtually impossible to incorporate D-amino acid into proteins without modifying the chemical structure of the peptidyl transferase center of naturally occurring tRNA, it is possible to achieve the corresponding D-L peptide bond formation in the model puromycin system. It may be noted that the alignment and surrounding environment of the amino terminals in the two cases are different. The attachment of the amino acid moieties at the CCA end of amino and peptidyl terminals restricts their free rotation. The limited available orientational degrees of freedom are, however, different in different systems studied: tRNA or puromycin or other analogues. Consequently, the degrees of freedom available and the extent of confinement do vary, and these factors affect the interaction and discrimination in the respective cases.

Starck et al. used a series of synthetic puromycin analogues to measure the activity of D-amino acid and β-amino acids in an intact eukaryotic translation system (rabbit reticulocyte ribosome) (Starck et al., 2003, 8090). The puromycin derivatives differ in amino acid stereochemistry and amino acid moiety as well as the number of carbon units in the amino acid backbone. The activity of the model compounds is measured in a high dynamic range IC_{50} potency assay using the rabbit reticulocyte protein synthesis system. While the L-puromycin inhibits globin mRNA translation with an IC_{50} value of 1.8 μm, the corresponding D-puromycin or 9-{3'-deoxy-3'-[(4-methoxy-D-phenylalanyl)amino]-β-D-ribofuranosyl}-6(N,N' dimethylamino) purine) inhibits globin translation, giving an IC_{50} value of 280 μm. The difference is 150-fold. The activity of a modified puromycin derivative (9-{3'-deoxy-3'-[(4-methyl-L-phenylalanyl) amino]-β-D-ribofuranosyl}-6(N, N'-dimethylamino) purine or L-(4-Me)-Phe-PANS) is found to be highest, giving an IC_{50} value of 1.0 μm. However, its D-amino acid isomer has an IC_{50} value of 2400 μm, which is even lower than D-puromycin and 2400-fold less potent than the corresponding L-isomer. Alanine analogues show little difference (~3-fold) between the L- and D-isomer. The size and geometry of the side chain is suggested to play an important role in the synthesis, with larger hydrophobic side chains having improved function. It is also indicated that the structural basis of the stereoselectivity could not be addressed due to the unavailability of the high-resolution structure of the rabbit reticulocyte ribosome. However, modeling of the placement of D-puromycin in the active site of *Haloarcula marismortui* 50S, for which the high-resolution structure is available, subunit is attempted. U2620 residue is found to be the closest nucleotide to the D-side chain. This view is supported by the structural analysis, which suggested that the steric hindrance of the molecular segments of the D-amino acid could lead to unfavorable incorporation of D-isomer. The role of the surrounding environment of the active site and particularly the role of U2620 in the discrimination in the peptide synthesis was recently studied using computational methods and will be discussed later.

While the foregoing studies indicated that D-amino acid incorporation into proteins is practically impossible, Hecht and coworkers pointed out that alteration of the peptidyltransferase center might lead to enhanced D-amino acid incorporation (Dedkova et al., 2003, 6616). Mutations in the 23SrRNA in the region of PTC and helix 89 lead to conformational change in ribosome that alters its behavior in the protein synthesis. Modified ribosomes with mutations in regions 2447–2450 (belongs to the PTC region) and 2457–2462 (belongs to helix 89 region) of *E. coli* 23S rRNA and cell-free protein-synthesizing systems were prepared from mutant ribosomes. A high level of suppression in the presence of D-methionine (23%) and D-phenylalanine (12%) is observed. This indicates that in ribosome, the active site structure is, at least partly, responsible for the discriminating mechanism against D-aminoacyl-tRNA$_{CUA}$'s in the ribosomal A site. Putative alterations may lead to enhanced incorporation of D-amino acid. The result indicates that the surrounding of PTC has a major influence on the process of synthesis. This result is supported by recent computational studies, where it is shown that the removal of a residue can reduce chiral discrimination; this will be discussed later.

The foregoing experimental observations show that chiral discrimination in peptide synthesis is significant. The incorporation of D-amino acid is unfavored in natural synthesis, but can be carried out in modified systems such as mutated active sites or puromycine. A few features of the reaction are understood from the experimental studies. First, the intermolecular energy profile of the reaction is expected to be orientation dependent. Consequently, chiral discrimination may not be averaged out for all possible orientational degrees of freedom as in the case of the same reaction in a solvent. Further, molecular segments involved in the reaction (L-amino acid and D-sugar) being chiral, their chirality should influence the nature of the corresponding energy surface. Confinement within a nanodimension must influence the discrimination exhibited by the reaction as also observed in biomimetic systems. It is pointed out in the literature that the preorganization of the active site residues and their proximity is important for effective catalysis during the progress of the reaction (Sharma et al., 2005, 11307). All these factors are recently explored using computational methods to understand the molecular mechanism of the discrimination.

Peptide bond formation was studied using *ab initio* methods for L-alanine dipeptide formation considering both a concerted and stepwise mechanism (Thirumoorthy and Nandi, 2007b, 107). In this study, the reacting amino acids are not covalently linked to any part of a larger structure, unlike as present in the CCA terminal in tRNA during peptide synthesis in PTC. The amino acid moieties in the model are not surrounded by the active site moieties. It is unlikely that chiral discrimination will be exhibited in such a system as the reacting moieties are free to assume various mutual arrangements without being constrained as they are not covalently linked to a larger structure (RNA) and neighboring residues (composing the PTC). As a result, discrimination, which is averaged out over all possible mutual orientations, is not expected to be observed. However, the reacting segments in PTC in tRNA are constrained as they are covalently linked with A- and P-terminals and are located at a distance of nanometer range. The A- and P-terminals are surrounded by nucleic acid residues, which are also within nanoscale separation. Consequently, significant and observable chiral discrimination in the reaction within the PTC is expected to be observed due to the influence of restricted rotation, confinement, and interaction with neighboring residues. These factors are important for chiral discrimination, as noted in biomimetic systems such as monolayers and bilayers discussed in the introduction. The orientation dependence of the reaction is also important for chiral discrimination for the following reason. During the peptide bond formation, the 3′ end of the A-site arrives at the P-site by a process of mutual rotation within the active site (Figure 3.3) (Zarivach et al., 2004, 901). The discrimination in the intermolecular energy surfaces of a pair of alanine molecules in the restricted orientational state (a model of the A- and P-site terminals during their restricted rotation in PTC) is investigated (Thirumoorthy and Nandi, 2006, 8840). Starting from the optimized structures of the nonbonded homochiral (L-L) and heterochiral (D-L) pairs of molecules, the energy surfaces are studied with rigid geometry by varying the distance and orientation. The molecular arrangement is shown in Figure 3.4. The potential energy surface is scanned with the rigid geometry of the residues as a function of distance and orientation starting from the optimized geometry.

The justification of such variation is that peptide bond formation is possible only when the corresponding -NH$_2$ group of the amino terminal is in the proper orientation suitable for the nucleophilic attack at the carbonyl carbon of the carboxyl group. As the orientation dependence of the proximity effect of the peptide bond formation is important, the variation in energy of the L-L and D-L pair with the dihedral angle will reveal the chiral discrimination of the corresponding pairs. The resulting energy profile is expected to provide information about the energetic advantage (if any) of the approach of an L-alanine (with varying orientation and distance) to another L-alanine compared to the corresponding process in which a D-alanine molecule approaches the L-alanine molecule.

The energy difference between the L-L-pair and D-L-pair at a given intermolecular orientation and separation for the same value of both variables for both pairs is defined as the chiral discrimination energy and is denoted by ΔE_{LL-DL}. The results are shown in Figure 3.5a–f. The difference in the nature of the energy surfaces of the respective pairs can be

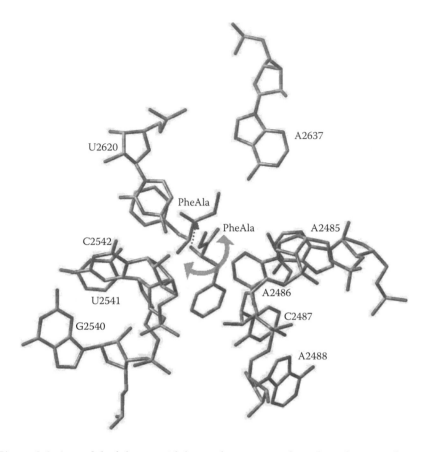

Figure 3.3 A model of the peptidyl transferase center based on the crystal structure of CCA-Phe-cap-biotin bound simultaneously at half-occupancy to both
the A-site and P-site of the 50S ribosomal subunit. The segments are taken from
the crystal structure of the ribosomal part of *Haloarcula marismortui*. (1Q86.PDB).
Residues include (all numbering corresponds to scheme as in *Haloarcula marismortui*): A76 of CCA attached to phenylalanine at A-terminal, A76 of CCA attached
to phenylalanine at P-terminal, A2485, A2486, C2487, A2488, U2620, A2637,
G2540, U2451, and C2452 (Thirumoorthy, K. and Nandi, N. 2007a. *J. Phys. Chem.
B*, 111: 9999. Reprinted with permission from the American Chemical Society. ©
American Chemical Society).

understood from $\Delta E_{LL\text{-}DL}$ as a function of orientation. Homochiral preference is observed when $\Delta E_{LL\text{-}DL}$ is negative, and when $\Delta E_{LL\text{-}DL}$ is positive, a heterochiral preference is noted. The discrimination is computed
for both the neutral and zwitterionic states. It is well known that BSSE
(basis set superposition error) gives rise to an artificial lowering in energy
in the computation of intermolecular interaction (van Duijneveldt, 1994,
1873), and it is ensured that the discrimination observed is not artificially

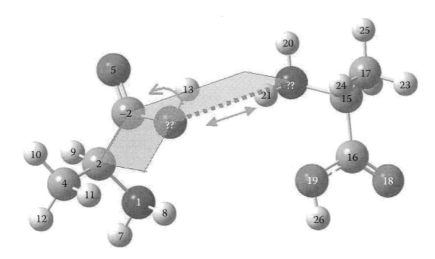

Figure 3.4 The scheme of mutual dihedral angle variation in a pair of nonbonded neutral L-L alanine molecules to study the discrimination in the orientation-dependent interaction (Thirumoorthy, K. and Nandi, N. 2006. *J. Phys. Chem. B.*, 110: 8840. Reprinted with permission from the American Chemical Society. © American Chemical Society).

generated by BSSE. The energy surface is calculated by MP2/6-311++G** level of theory. BSSE is further corrected at the same (MP2/6-311++G**) level using the counterpoise method. The energy profile of the naturally available achiral amino acid, glycine in zwitterionic state, is also investigated to better compare the effect of chirality of the alanine. The orientation and distance dependence is investigated as described before for a pair of alanine molecules.

The result shows that the interaction energy of the L-L pair and the corresponding interaction energy of the D-L pair is not identical with either variation of distance or with variation of the mutual orientation of the two molecules. The potential energy surfaces of the L-L and D-L pairs are found to be dissimilar and the reflect the underlying chirality of the homochiral pair and the racemic nature of the heterochiral pair. The plot of the L-L pair is dissymmetric, revealing the underlying chirality of the molecules. On the contrary, the plot for the D-L pair is symmetric around a 180° orientation, which reveals the racemic state of the neutral pair where the charge dissymmetry is also absent. Unlike the neutral case, the asymmetric charge distribution of the neighboring molecules (starting geometry) breaks the symmetry of the energy profile of the D-L zwitterionic pair, which can be observed in the corresponding dissymmetric nature of the energy profile. The conformational energy differences between the D or L molecules themselves are negligibly small (the energies are identical for their optimized structures), and the observed difference in the

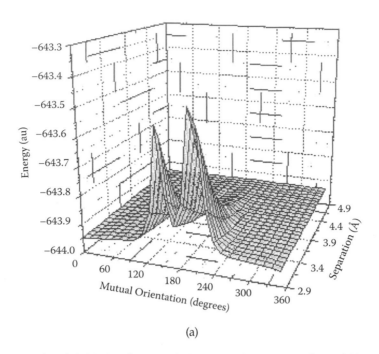

(a)

Figure 3.5 The *ab initio* (HF/6-311++G**) potential energy surface of (a) a pair of neutral L-L alanine molecules (Thirumoorthy, K. and Nandi, N. 2006. *J. Phys. Chem. B.*, 110: 8840 Reprinted with permission from the American Chemical Society. © American Chemical Society). All calculations are carried out with rigid geometry starting from the optimized structure of the corresponding pair using Gaussian 03W suite of programs (Frisch, M.J. et al., 2004. *Gaussian 03, Revision C.02*, Wallingford, CT: Gaussian, Inc.). (continued)

intermolecular potential (ΔE_{LL-DL}) is not due to any conformational difference. The deviation in energies for the zwitterionic case between the L-L pair and D-L pair is due to the difference in the position of protons between the amide groups and carboxyl group in the respective cases.

It has been pointed out that the very large difference between the energy surfaces of the L-L and D-L pairs at specific orientations at the same distances between the corresponding pairs is due to short-range repulsive interactions. In general, the short-range repulsive part of both van der Waals and electrostatic (Coulombic, dipolar, or multipolar) interaction varies more sharply than the corresponding long-range attractive part of the intermolecular interaction (Israelachvili, 1985; Maitland et al., 1981). The energy profile of L-L at a given short-range separation is less unfavorable than the corresponding energy profile of the D-L pair at the same intermolecular separation. However, the variations in intermolecular interaction energy are large at such short-range intermolecular distances, and this causes the large discrimination observed (which is the

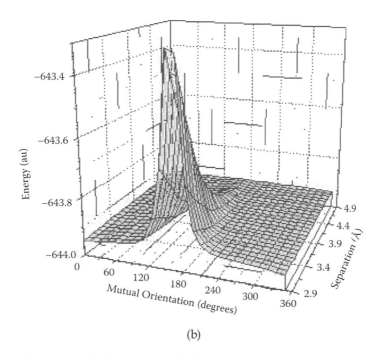

(b)

Figure 3.5 (continued) The *ab initio* (HF/6-311++G**) potential energy surface of (b) a pair of neutral D-L alanine molecules as a function of orientation and distance (Thirumoorthy, K. and Nandi, N. 2006. *J. Phys. Chem. B.,* 110: 8840 Reprinted with permission from the American Chemical Society. © American Chemical Society). All calculations are carried out with rigid geometry starting from the optimized structure of the corresponding pair using Gaussian 03W suite of programs (Frisch, M.J. et al., 2004. *Gaussian 03, Revision C.02,* Wallingford, CT: Gaussian, Inc.).

major source of the discrimination for neutral molecules). The favorable electrostatic interaction further contributes to the large homochirality of the ΔE_{LL-DL} for the zwitterionic pair in addition to the short-range steric repulsion. The basis set superposition error (BSSE)-corrected results show enhanced discrimination. Use of the higher-level Møller–Plesset perturbation theory (MP2) and further BSSE correction do not change the conclusions made at the Hartree–Fock (HF) level (Figure 3.6a and 3.6b, respectively). It is seen that BSSE correction enhances the discrimination at the HF level. ΔE_{LL-DL} was also calculated using MP2/6-311++G** theory and further corrected the BSSE using the same level of theory.

The major conclusions based on HF- and MP2-level calculations agree with the results from the B3LYP/6-311++G** level of density functional theory (DFT) calculations. The stable zwitterionic form of amino acid could not be obtained using DFT. The zwitterionic species can be stabilized by

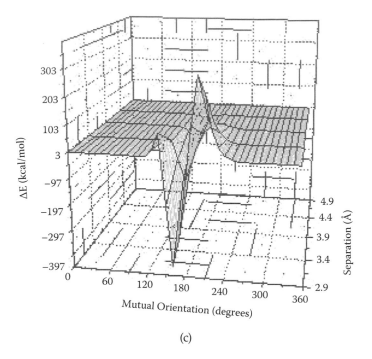

(c)

Figure 3.5 (continued) The *ab initio* (HF/6-311++G**) potential energy surface of (c) the chiral discrimination energy ($\Delta E_{LL\text{-}DL}$) calculated from the potential energy surface of pairs of neutral L-L and D-L alanine molecules (Thirumoorthy, K. and Nandi, N. 2006. *J. Phys. Chem. B.*, 110: 8840 Reprinted with permission from the American Chemical Society. © American Chemical Society). All calculations are carried out with rigid geometry starting from the optimized structure of the corresponding pair using Gaussian 03W suite of programs (Frisch, M.J. et al., 2004. *Gaussian 03, Revision C.02*, Wallingford, CT: Gaussian, Inc.).

incorporation of the solvation effect (Jalkanen et al., 2001, 125; Friman et al., 2000, 165; Knapp-Mohammady et al., 1999, 63; Tajkhorshid et al., 1998, 5899). However, when a pair of zwitterionic alanine molecules is considered, a given zwitterionic alanine molecule does not get converted into neutral species due to the presence of the carboxyl group of the neighboring molecule close to the proton attached to the amine group of a given alanine molecule. Besides, there are other reported limitations of the DFT in calculating the intermolecular interactions in general. It is known that since the functional form of the exact exchange-correlation energy is not known, current DFT methods cannot yield exact results. The lack of a systematic way to arrive at the exact result using a series of calculations is pointed out to be the deficiency of DFT methods. Also, the theory does not allow an evaluation in the errors for a calculation when the results converge with the increase in basis sets and the quality of a result can only be

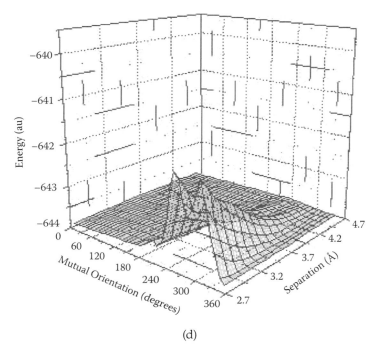

(d)

Figure 3.5 (continued) The *ab initio* (HF/6-311++G**) potential energy surface of (d) a pair of zwitterionic L-L alanine molecules (Thirumoorthy, K. and Nandi, N. 2006 *J. Phys. Chem. B.*, 110: 8840. Reprinted with permission from the American Chemical Society. © American Chemical Society). All calculations are carried out with rigid geometry starting from the optimized structure of the corresponding pair using Gaussian 03W suite of programs (Frisch, M.J. et al., 2004. *Gaussian 03, Revision C.02*, Wallingford, CT: Gaussian, Inc.).

verified by comparing with the experimental result or with a high-quality wave mechanical result (Jensen, 1999, 192). Despite these limitations, a number of successful DFT studies on the calculation of intermolecular interaction have been reported. (Zhang and Yang, 2005, 77; Chis et al., 2005, 153; Podeszwa and Szalewicz, 2005, 488; Williams and Chabalowski, 2001, 646). Notably, the cost of DFT calculations is only N^3, where N is the number of basis sets, while the costs for MP2 and MP4 are N^5 and N^7, respectively. Also, gradient-corrected results are either close or in some cases better than HF or MP2 results (Novoa and Sosa, 1995, 15837).

The chiral discrimination exhibited by a pair of alanines arises when their mutual rotatory path is orientationally restricted. Such discrimination in the intermolecular interaction is relevant to the process of peptide biosynthesis where the A- and P- terminals undergo restricted rotation during their path of approach to carry out the reaction. It is obvious that there would be little or no energetic advantage of L-L synthesis over

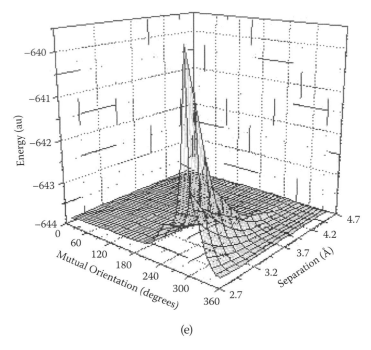

(e)

Figure 3.5 (continued) The *ab initio* (HF/6-311++G**) potential energy surface of (e) a pair of zwitterionic D-L alanine molecules as a function of orientation and distance (Thirumoorthy, K. and Nandi, N. 2006. *J. Phys. Chem. B.*, 110: 8840 Reprinted with permission from the American Chemical Society. © American Chemical Society). All calculations are carried out with rigid geometry starting from the optimized structure of the corresponding pair using Gaussian 03W suite of programs (Frisch, M.J. et al., 2004. *Gaussian 03, Revision C.02*, Wallingford, CT: Gaussian, Inc.).

D-L synthesis provided the interaction energy profile for the rotational motion of A to P site is independent of the mutual orientation of the concerned chiral species. The chiral discrimination observed could influence the preferred L-amino acid incorporation. It is interesting to note that the maximal value *of* $\Delta E_{\text{LL-DL}}^{\text{Homochiral}}$ (observed homochiral preference*)* is larger than the maximal value of $\Delta E_{\text{LL-DL}}^{\text{Heterochiral}}$ in the same plot (observed heterochiral preference) for both neutral and zwitterionic pairs of alanine using both the HF and the DFT level of theoretical calculations. Short-range steric overlap in the case of the D-L pair is a major reason for the preferred homochirality, which does not occur for the L-L pair. The electrostatic interaction at close separation further augments the homochirality in the case of the zwitterionic pair.

It is observed in the plot of $\Delta E_{\text{LL-DL}}$ for zwitterionic alanine pairs that the largest discrimination is not at the orientation corresponding to the

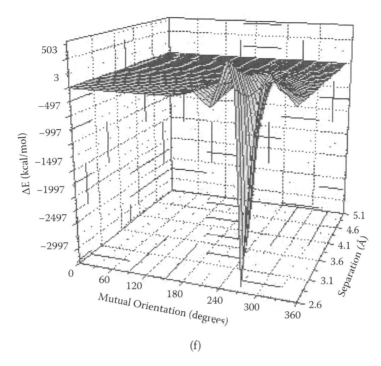

(f)

Figure 3.5 (continued) The *ab initio* (HF/6-311++G**) potential energy surface of (f) the chiral discrimination energy (ΔE_{LL-DL}) calculated from the potential energy surface of pairs of zwitterionic L-L and D-L alanine molecules (Thirumoorthy, K. and Nandi, N. 2006. *J. Phys. Chem. B.*, 110: 8840 Reprinted with permission from the American Chemical Society. © American Chemical Society). All calculations are carried out with rigid geometry starting from the optimized structure of the corresponding pair using Gaussian 03W suite of programs (Frisch, M.J. et al., 2004. *Gaussian 03, Revision C.02*, Wallingford, CT: Gaussian, Inc.).

optimized geometry of the respective pairs as in neutral alanine. The increase in energy due to short-range steric repulsion is highest when the hydrogen atom of the second amino group of D-alanine comes in steric hindrance with the oxygen atom of the second carboxylic group of L-alanine. This is due to the strong short-range overlap of atoms in the zwitterionic D-L pair. In the case of the L-L zwitterionic pair, on the other hand, the second carboxyl of the L molecule and second amine of the D-molecule are in close proximity at the same relative orientation. This gives rise to the short-range electrostatic attraction. Thus, the favorable electrostatic interaction further contributes to the large homochirality of the ΔE_{LL-DL} for the zwitterionic pair in addition to the short-range steric repulsion (which is the major source of the discrimination for neutral molecules).

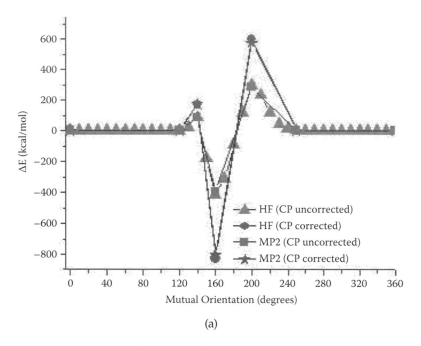

(a)

Figure 3.6 Comparison of the counterpoise-corrected result that is BSSE free (denoted by CP$_{Corrected}$) and uncorrected result (denoted by CP$_{Uncorrected}$) of the slice of the chiral discrimination energy for (a) neutral alanine (Thirumoorthy, K. and Nandi, N. 2006. *J. Phys. Chem. B.*, 110: 8840. Reprinted with permission from the American Chemical Society. © American Chemical Society). All calculations were performed by Gaussian 03W suite of programs (Frisch, M.J. et al., 2004. *Gaussian 03, Revision C.02*, Wallingford, CT: Gaussian, Inc.). (continued)

Experimental studies have indicated that the 3′ ends of the A- and P-site tRNA have to face each other in order to form the proper stereo-chemistry related to peptide bond formation. A process of spiral rotation of the 3′ end of the tRNA achieves this. The outcome of this spiral motion is that a P-site carbonyl carbon atom faces the A-site nucleophilic amine. This occurs without any significant conformational alterations of the 3′ end. During the A-P rotatory motion, the L-L and D-L pairs of A- and P-terminals have to pass through orientations that are different from the intermolecular mutual orientation corresponding to the optimized geom-etry of the pair of molecules (while the individual molecules may remain in their respective optimized geometries). This passage is easier for the L-L pair rather than the D-L pair. In other words, during the A-P rotatory movement, the L-L pair will pass through relatively low-energy regions of the intermolecular energy surface compared to the D-L pair. The inter-molecular energy surface of the L-L pair is more favorable than the cor-responding energy surface of the D-L pair. The study suggests a possible

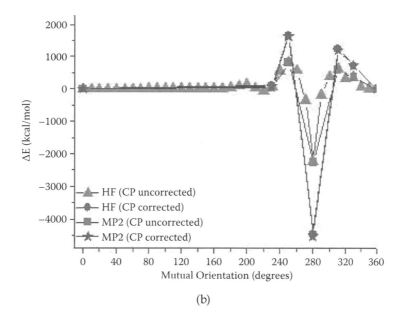

(b)

Figure 3.6 (continued) comparison of the counterpoise-corrected result that is BSSE free (denoted by CP$_{Corrected}$) and uncorrected result (denoted by CP$_{Uncorrected}$) of the slice of the chiral discrimination energy for (b) zwitterionic alanine molecules. The plots correspond to the distance at which the maximal homochirality and heterochirality are observed in the respective uncorrected energy surfaces. Similarly, the line with filled squares indicates the energy plot for the BSSE-uncorrected profile using chiral discrimination energy values at the MP2/6-311++G**. The line with asterisks indicates the energy plot for the BSSE-corrected profile of chiral discrimination energy values at the MP2/6-311++G** (Thirumoorthy, K. and Nandi, N. 2006. *J. Phys. Chem. B.*, 110: 8840. Reprinted with permission from the American Chemical Society. © American Chemical Society). All calculations were performed by Gaussian 03W suite of programs (Frisch, M.J. et al., 2004. *Gaussian 03, Revision C.02*, Wallingford, CT: Gaussian, Inc.).

mechanism of D-amino acid exclusion. This is due to the larger degree of steric hindrance between D- and L-amino acid themselves than the L-L pair and concomitant homochiral preference. The electrostatic interaction of the homochiral pair further augments the homochirality in the case of the zwitterionic pair. The result indicates that chiral discrimination in peptide synthesis can arise, at least partly, from the interaction between amino acids themselves as a function of intermolecular separation and orientation when the two amino acids are restricted to follow a rotatory path as occurs during peptide synthesis when A- and P-terminals are covalently linked to the tRNA structure and are not allowed to assume all possible mutual orientations. The result can also be correlated with the observed discrimination noted in model systems such as puromycine, where the amino acids undergo restricted rotation but are not confined by surrounding residues.

The effect of the surroundings of the PTC on chiral discrimination is studied in a model of peptidyl transferase center from the crystal structure of *Haloarcula marismortui* using hybrid quantum chemical studies (Thirumoorthy and Nandi, 2007, 9999). The crystal structure of the CCA-Phe-cap-biotin bound simultaneously at half-occupancy to both the A-site and P-site of the 50S ribosomal subunit (Hansen et al., 2002, 11670) is used to generate the molecular segments used in the theoretical calculation. Residues located in close proximity to A- and P- terminal during rotation are as follows (all numbers correspond to the scheme given in *Haloarcula marismortui*): phenyl alanine at A-terminal, phenyl alanine at P-terminal, A2485, A2486, C2487, A2488, U2620, A2637, G2540, U2451, and C2452. The model is shown in Figure 3.3. Phenyl alanine at the A-terminal has either L-(natural) or D-configuration and phenyl alanine at the P-terminal has the L-configuration in calculations.

The study shows that the interaction of the L-L pair can happen at a significantly lower energy and without any steric constraint over the range of orientation as shown in Figure 3.7. The interaction between the D-L pairs is relatively unfavorable, and the corresponding energy profile has two minima only at a limited range of orientations. The orientational space for the D-L pair is limited; this pair experiences more steric clash with surrounding residues compared to the L-L pair. This is clear evidence of homochiral preference. In order to understand the origin of the observed discrimination, different surrounding residues are removed and the influence on the ease of the rotatory path is studied. Exclusion of the U2620 from the surrounding residues drastically diminishes the discrimination. This is shown in Figure 3.8a. This indicates that the U2620 has preferential nonbonded interaction with A- and P- terminals in the L-L form rather than the D-L form. However, homochiral preference is still significant, as shown in Figure 3.8b, where the scale is enlarged. The entire preferred homochirality, as observed in the experiment and calculated as in Figure 3.7, is not due to U2620, and the residual discrimination is nonnegligible. All other residues are successively removed (A2485, A2486, C2487, A2488, A2637, G2540, U2451, and C2452) in order to identify the origin of residual discrimination, and this is noted to change little from that observed in Figure 3.8b. It may be noted that the removal of backside residues such as A2485, A 2486, C2487, and A2488 gradually diminishes the repulsive steric constraint, but discrimination remained effectively unchanged from that observed in Figure 3.7. Finally, all residues are removed, and two phenyl alanine groups are rotated without any surrounding residues in the same specified range of rotation. Interestingly, the discrimination is nonvanishing, as shown in Figure 3.9. A strong

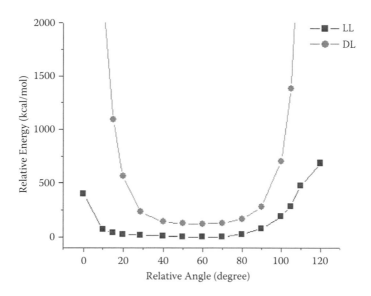

Figure 3.7 The variation in interaction energy as a function of orientation of phenyl alanine at A-terminal and phenyl alanine at P-terminal in the presence of A2485, A2486, C2487, A2488, U2620, A2637, G2540, U2451, and C2452 residues as shown in Figure 3.2 for L-L and D-L pair combinations of two terminals subunit. Starting from the mutual arrangement of A- and P- terminals in crystal structure, the A-terminal is oriented clockwise by 60° and anticlockwise by an angle 60° with 10° increments covering a total range of orientation of 120° (Thirumoorthy, K. and Nandi, N. 2007a. *J. Phys. Chem. B*, 111: 9999. Reprinted with permission from the American Chemical Society. © American Chemical Society). All calculations were performed by Gaussian 03W suite of programs (Frisch, M.J. et al., 2004. *Gaussian 03, Revision C.02*, Wallingford, CT: Gaussian, Inc.).

homochiral preference is still noted, which is entirely determined by the amino acid side chains. Comparison of the relative energy gap of L-L and D-L pair plots, including all surrounding residues except U2620 and the relative energy gap of the L-L and D-L pairs with all surrounding residues removed clearly shows that the residual discrimination (remaining after removal of U2620) remained the same in the specified range of orientation and is independent of other residues (Figure 3.10).

This conclusion corroborates the results of a previous study based on *ab initio* and DFT studies on interaction of alanine molecules where homochiral preference was noted to be a function of mutual orientation (Thirumoorthy and Nandi, 2006, 8840). The study quantitatively shows that the observed homochiral preference is due to U2620 residue as well as the amino acid side chain at the A- and P-terminals. A major part of the discrimination comes from the variation of nonbonded interaction of rotating A-terminal with U2620 during the approach of the former

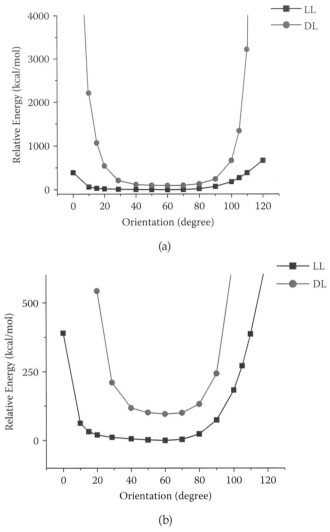

(a)

(b)

Figure 3.8 (a) The variation in interaction energy as a function of orientation of phenyl alanine at the A-terminal and phenyl alanine at the P-terminal in the presence of A2485, A2486, C2487, A2488, A2637, G2540, U2451, and C2452 residues and excluding U2620 for L-L and D-L pair combinations of two terminals subunit. Starting from the mutual arrangement of A- and P- terminals in crystal structure, the A-terminal is oriented clockwise by 60° and anticlockwise by an angle of 60° with 10° increments covering a total range of orientation as 120°. (b) Enlarged image of the discrimination remaining after removal of U2620 (Thirumoorthy, K. and Nandi, N. 2007a. *J. Phys. Chem. B*, 111: 9999.) (Reprinted with permission from the American Chemical Society. © American Chemical Society). All calculations were performed by Gaussian 03W suite of programs (Frisch, M.J. et al., 2004. *Gaussian 03, Revision C.02*, Wallingford, CT: Gaussian, Inc.).

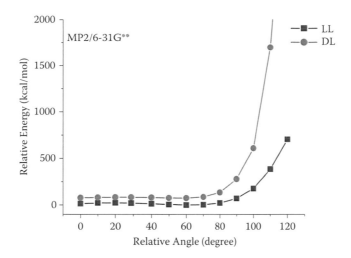

Figure 3.9 The variation in interaction energy as a function of orientation of phenyl alanine at the A-terminal and phenyl alanine at the P-terminal for L-L and D-L pair combinations of two terminals without any surrounding residue subunit. Starting from the mutual arrangement of A- and P- terminals in crystal structure, the A-terminal is oriented clockwise by 60° and anticlockwise by an angle of 60° with 10° increments covering a total range of orientation as 120° (Thirumoorthy, K. and Nandi, N. 2007a. *J. Phys. Chem. B*, 111: 9999. Reprinted with permission from the American Chemical Society. © American Chemical Society). All calculations were performed by Gaussian 03W suite of programs (Frisch, M J et al., 2004. *Gaussian 03, Revision C.02*, Wallingford, CT: Gaussian, Inc.).

toward the P- terminal. Significant discrimination is due to the difference in the side chain interaction of A- and P-terminals themselves during the rotatory motion even when no surrounding residues are present.

It is important to address how important is the chirality of the D-sugar in the context of the long-length-scale structural organization of the PTC. Explicitly, can the possible heteropairs and homopairs (other than the D-sugar and L-amino acid) perform peptide synthesis with similar efficiency in the natural biological surroundings (PTC)? As the enantiomers differ only in the mutual spatial arrangement of groups, changes in the spatial interactions in other hetero- and homopair combinations within the PTC may be negligible for discrimination. The question is whether such changes in spatial orientation have any influence on the effective peptide synthesis. Several complementarities of the sugar chirality with amino acid and the surrounding bases can exist that might make favorable contribution of the D-sugar ring to the process of chiral discrimination in the peptide synthesis. As mentioned in the introduction, a molecular understanding of the origin of the chiral specificity of the sugar ring (D-form rather than the L-form) in nucleic acid structure and

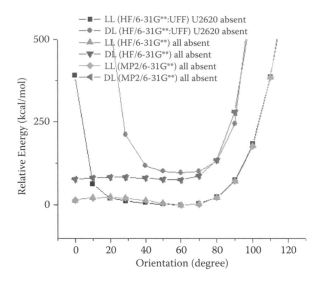

Figure 3.10 Comparison of the relative energy gap of L-L and D-L pairs including all surrounding residues except U2620 and the relative energy gap of L-L and D-L pairs with all surrounding residues removed (Thirumoorthy, K. and Nandi, N. 2007a. *J. Phys. Chem. B,* 111: 9999. Reprinted with permission from the American Chemical Society. © American Chemical Society). All calculations were performed by Gaussian 03W suite of programs (Frisch, M.J. et al., 2004. *Gaussian 03, Revision C.02,* Wallingford, CT: Gaussian, Inc.).

its role in controlling the fidelity of the peptide synthesis is important. Experimental studies of Hecht and coworkers (Hecht, 1992, 545) indicated that while the *N*-acetyl-D-phenylalanyl-tRNA[Phe] and *N*-acetyl-D-tyrosyl-tRNA[Phe] produced only a small amount of dipeptide, the 2′ and 3′ deoxyadenosine analogues produced no detectable dipeptide.

The possibility that the discrimination in peptide synthesis might be coupled with the D-form of sugar was investigated recently (Thirumoorthy and Nandi, 2008, 9187). First, it is possible that the D-form of sugar ring might have favorable stereochemistry related to the structural organization of the tRNA itself. Second, the sugar ring might have the proper stereochemistry that can make the rotatory path (through which the A-site tRNA 3′-end flips into the P-site by rotating around the bond connecting the single strand 3′-end) leading to the optimal orientation for peptide bond formation easier to achieve. Third, the hydroxyl group attached to 2′ can catalytically lower the transition state barrier (related to the formation of the peptide bond) when positioned in a favorable stereochemical orientation.

Modeling studies show that the alterations in the chiral centers of the ribose sugar ring bring large-scale structural changes in the tRNA structure. Such structural modifications are unfavorable for the peptide synthesis process (Thirumoorthy and Nandi, 2008, 9187). Model building indicates that the unnatural combinations of amino acid and sugar (L-amino acid: L-sugar, D-amino acid: D-sugar, and D-amino acid: L-sugar) are unfavorable for the proper stereochemistry needed to form a peptide bond within the PTC as well as creating steric hindrance. Large-scale structural rearrangement is required to accommodate these unnatural pairs (Thirumoorthy and Nandi, 2008, 9187). Exchanging the C_5' with any other atom or group leads to a change in the orientation of the phosphodiester backbone. This is energetically unfavorable. Similarly, a change in the chirality at C_1' will lead to a change in the orientation of bases and will affect the base-pairing. The result indicates that naturally available PTC is highly specific about the chiralities of the amino acid and sugar ring and can accommodate only the natural heterochiral pair of sugar and amino acid without any large-scale structural change in tRNA.

The influence of the ribose sugar ring on the rotatory path of the approach of the two terminals toward each other has also been studied (Thirumoorthy and Nandi, 2008, 9187). The result shows that the presence of a sugar ring in both terminals makes the rotatory path more favorable compared to the case when the sugar ring is removed (Figure 3.11). Since rotation of the A-terminal occurs around the 3' bond, the rotatory process is a mutual orientational motion of the amino terminal with respect to the A-site sugar ring. As a result, removal of the A-site sugar ring raises the energy due to the loss of favorable interaction. The contribution of the P-site sugar ring in stabilizing the process is more than that of the A-site sugar ring. This result is consistent with the fact that as the A-site approaches the P-site, the amino group of the A-site interacts with the peptidyl terminal and sugar ring of the P-site. This contributes to the orientation-dependent interaction that is lost with the removal of the ring, and the energy increases further. The result shows that the A- and P-site sugar rings have a significant favorable influence on the rotatory path. It further indicates that the combined presence of D-sugar and L-amino acid makes the rotatory path more favorable compared to the structure in which the D-sugar is absent.

The ribose sugar ring also has another influence on the peptide bond formation process. The role of the microscopic chirality of the sugar ring and specifically that of the 2' chiral carbon of the A76 seems important. Its catalytic role in peptide synthesis is studied (Thirumoorthy and Nandi, 2008, 9187). Experimental studies indicate that the hydroxyl group is responsible for the significantly high rate of peptide synthesis in ribosomal catalysis compared to that of an uncatalyzed one (in the absence of ribosome). The catalytic mechanism of the involvement of the 2' OH group of the sugar ring remained a question. The role of 2' OH in the peptide synthesis has been suggested for a long time (Chamberlin et

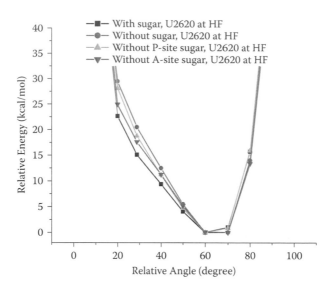

Figure 3.11 The variation in interaction energy as a function of orientation of phenyl alanine at the A-terminal and phenyl alanine at the P-terminal in the presence and in the absence of sugars attached to A-76 attached to both terminals and within a model of peptidyl transferase center Starting from the mutual arrangement of A- and P- terminals in crystal structure, the A-terminal is oriented clockwise by 60° and anticlockwise by an angle 60° with 10° increments covering a total range of orientation of 120°. The ONIOM calculation is performed at (HF/6-31G**: UFF). The phenyl alanine at the A-terminal, P-terminal, and U2620 is considered to be at the QM level. The residues other than U2620 included in the model are considered to be at the UFF level. The variation in interaction energy as a function of orientation of phenyl alanine at the A-terminal and phenyl alanine at the P-terminal in the presence and in the absence of sugars of only the A-site and P-site, respectively are also shown (Thirumoorthy, K. and Nandi, N. 2008. *J. Phys. Chem. B.*, 112: 9187. Reprinted with permission from the American Chemical Society. © American Chemical Society). All calculations were performed by Gaussian 03W suite of programs (Frisch, M.J. et al., 2004. *Gaussian 03, Revision C.02*, Wallingford, CT: Gaussian, Inc.).

al., 2002, 14688). A comparison of the activities of 2′ deoxy derivative of AcLeuAMP, an AcLeuAMP derivative with either a second AcLeu at the 2′ OH or with a 2′OCH$_3$ at the 2′ position of ribose, indicates that the activity of the deoxy substrate is at least 100-fold lower than the substrates containing active 2′ hydroxyl group (Dorner et al., 2002, 1131). Substitution of the P-site tRNA A76 2′ OH with 2′ H or 2′ F resulted in a 10^6-fold decrease in the peptide bond formation (Weinger et al., 2004, 1101). It is noted that the rate effect is inconsistent with the pH-dependent catalytic mechanism, and the role of the 2′OH could be orienting the nucleophile, stabilizing the transition state, or inducing a favorable catalytic conformation of the PTC

(Rodnina et al., 2005, 493). It has been proposed that it facilitates peptide bond formation by substrate positioning and acting as a proton shuttle between the amino group of A-site tRNA and the A76 3′ oxygen atom of the ester bond of the peptidyl-tRNA. A study of modified substrates and mimics of intermediates indicates that 2′OH can serve as a proton shuttle (Weinger and Strobel, 2007, 110). A proton shuttle is proposed in which the 2′OH receives a proton from the attacking amino group and simultaneously donates the proton to the neighboring 3′OH as the transition state is decomposed into product (Trobro and Åqvist, 2005, 12395). It is proposed that the catalytic effect is due to the stable hydrogen bond network observed along the reaction. However, other catalytic pathways involving 2′OH are possible. First, it is possible that the amino group attacks the ester carbon to yield the tetrahedral intermediate, which breaks down to deacylated tRNA and elongated peptidyl-tRNA. Second, it is possible that the zwitterionic intermediate may break down through a proton shuttle via the 2′OH of A76 in the P site stepwise or in a concerted pathway and, finally, the proton shuttle may involve a water molecule that interacts with 2′ and 3′hydroxyl of A76 (Schmeing et al., 2005, 437).

A recent detailed quantum mechanical calculation of the transition state also emphasized the role of increase in the hydrogen bonding between the transition state geometry and the ribosomal components in stabilizing the transition state (Gindulyte et al., 2006, 13327). It was suggested that the 2′OH group of the sugar ring remains in close interaction with the A-site amine and carbonyl group during the rotatory path leading to peptide bond formation through substrate-assisted catalysis. This mechanism is different from the proton shuttle mechanism. The role of the 2′OH group in the rate enhancement produced by ribosome based on entropic contribution was proposed (Sievers et al., 2004, 7897, Schweet and Allen, 1958, 1104). These studies indicate that the stereochemistry of the 2′ chiral carbon facilitates the reaction by catalytic mechanism using the 2′OH group despite the fact that the two mechanisms are found to be competitive.

The barrier heights of the transition states for the formation of the peptide bond and the influence of the 2′OH group are studied for two different mechanisms (proton shuttle and anchoring mechanism), and the difference in the barrier heights is attributed to the absence of the 2′OH group, which results in removal of the chirality of the 2′ carbon of the sugar ring (Thirumoorthy and Nandi, 2008, 9187). The transition state geometry of the reaction in the absence of the 2′OH of the sugar ring and the respective transition state geometry when 2′OH is present (as in the natural form of the sugar ring) are shown in Figures 3.12a and 3.12b, respectively. The geometries of the transition states are calculated from the models built from ribosomal parts of *Haloarcula marismortui* using HF level of theory (Thirumoorthy and Nandi, 2008, 9187) and from DFT theory based on a model from the crystal structure of 50S large

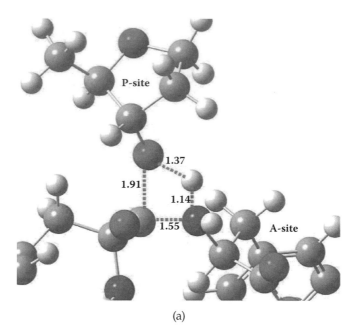

(a)

Figure 3.12 Comparison of the transition state geometries of L-L phenyl ala-nine peptide bond formation reaction (a) without 2′ OH of the sugar rings at the HF/3-21G level of theory (Thirumoorthy, K. and Nandi, N. 2008. *J. Phys. Chem. B.*, 112: 9187. Reprinted with permission from the American Chemical Society. © American Chemical Society). All calculations were performed by Gaussian (Frisch, M.J. et al., 2004. *Gaussian 03, Revision C.02*, Wallingford, CT: Gaussian, Inc.). (continued)

ribosomal subunit from *Deinococcus radiodurans* complexed with a tRNA acceptor stem mimic (Gindulyte et al., 2006, 13327). Although the theoreti-cal methods employed and the species studied are different (*Haloarcula marismortui* and *Deinococcus radiodurans*), the results obtained show no sig-nificant differences in the mechanisms proposed in the respective cases (Thirumoorthy and Nandi, 2008, 9187). It is noted that the reactant geom-etry contains a cycle of hydrogen bonds, and the hydrogen bond distances are further shortened in the transition state geometry. Several of these hydrogen bonds are absent when the 2′ OH is removed. The correspond-ing transition state (in absence of the 2′ OH) contains a hydrogen bond of distance 1.37 Å between the A-site α-amine hydrogen and P-site 3′ oxy-gen atom of the ester bond. The result indicates that all catalyzing bonds present in the transition state structure containing the P-site 2′ OH are missing in the corresponding transition state structure when the 2′ center is made achiral (Figure 3.12a). While the proton shuttle and anchoring mechanisms differ substantially, both lower the transition state energy

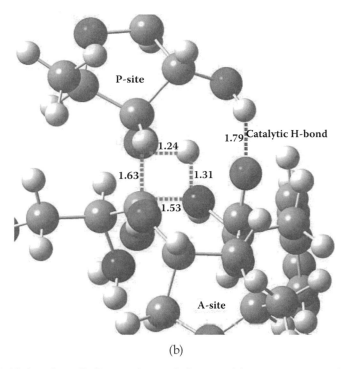

(b)

Figure 3.12 (continued) Comparison of the transition state geometries of L-L phenyl alanine peptide bond formation reaction (b) with the 2′ OH of the sugar rings at the HF/3-21G level of theory (Thirumoorthy, K. and Nandi, N. 2008. *J. Phys. Chem. B.*, 112: 9187. Reprinted with permission from the American Chemical Society. © American Chemical Society). All calculations were performed by Gaussian (Frisch, M.J. et al., 2004. *Gaussian 03, Revision C.02*, Wallingford, CT: Gaussian, Inc.).

compared to the case when the chirality of the 2′ carbon is either altered or removed. This can be followed from the geometries of the corresponding transition states (where the OH group is taking part in a proton shuttle and where the OH group acts as an anchor). Both geometries are shown in Figures 3.13a and 3.13b, respectively.

The result qualitatively indicates that the change in the stereochemistry of the 2′ center can alter the progress of the synthesis by several orders of magnitude. The observation corroborates the experimental data that the substitution of P- site tRNA 2′ OH by 2′ H and 2′ F results in a 10^6-fold reduction in the rate of peptide bond formation (Weinger et al., 2004, 1101). A comparison of the transition states in the two cases indicates that the anchoring role of the 2′ OH group is responsible for lowering the barrier height compared to the case when the 2′ OH group is removed. The result strongly suggests that the chirality of the 2′ carbon is vital for the rate enhancement.

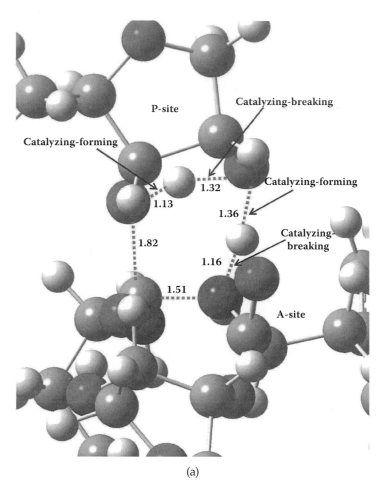

(a)

Figure 3.13 Comparison of the transition state geometries of L-L phenyl alanine peptide bond formation reaction with 2′ OH of the sugar ring with (a) proton shuttle (Thirumoorthy, K. and Nandi, N. 2008. *J. Phys. Chem. B.*, 112: 9187. Reprinted with permission from the American Chemical Society. © American Chemical Society). All calculations were performed by Gaussian 03W suite of programs (Frisch, M.J. et al., 2004. *Gaussian 03, Revision C.02*, Wallingford, CT: Gaussian, Inc.). (continued)

If the proper orientation of the 2′ OH and its role as an anchor in the transition state is responsible for the effective catalysis, then it is possible that the change in the orientation of the group will lead to a diminishing rate. Alteration of the stereochemistry at the 2′ center of the sugar ring results in increase of the barrier height. The result indicates that the change in the stereochemistry of the 2′ OH makes the process unfavorable. The rate is diminished less than when the OH group is completely

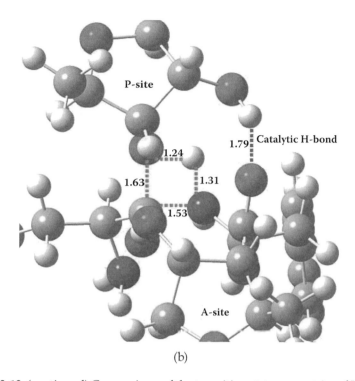

(b)

Figure 3.13 (continued) Comparison of the transition state geometries of L-L phenyl alanine peptide bond formation reaction with 2′ OH of the sugar ring with (b) anchoring mechanism at the HF/3-21G level of theory (Thirumoorthy, K. and Nandi, N. 2008. *J. Phys. Chem. B.*, 112: 9187. Reprinted with permission from the American Chemical Society. © American Chemical Society). All calculations were performed by Gaussian 03W suite of programs (Frisch, M.J. et al., 2004. *Gaussian 03, Revision C.02*, Wallingford, CT: Gaussian, Inc.).

removed. This indicates that the proper positioning of the group as well as the natural chirality (with respect to the 2′ center, at least) is vital for the reaction. The stereochemical orientation of the hydroxyl group is changed when the chirality at 2′ carbon is either altered or removed. This affects the proximal geometry for the catalytic process and leads to the reduction in the rate of peptide synthesis, which is consistent with experimental studies (Weinger et al., 2004, 1101).

The surrounding residues present in the PTC (which are in nanometer-length-scale proximity of the D-sugar) also influence the rotatory path and the discrimination. Although the bases are achiral, their interaction with the sugar ring can be orientation dependent and can influence chiral discrimination. It was pointed out earlier that when achiral moieties have restricted orientational freedom, their interaction profile with a chiral moiety can have orientation dependence (Nandi,

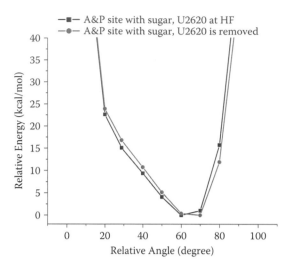

Figure 3.14 Variation in interaction energy as a function of orientation of phenyl alanine at the A-terminal and phenyl alanine at the P-terminal in the presence and in the absence of U2620, respectively (Thirumoorthy, K. and Nandi, N. 2008. *J. Phys. Chem. B.*, 112: 9187. Reprinted with permission from the American Chemical Society. © American Chemical Society). All calculations were performed by Gaussian 03W suite of programs (Frisch, M.J. et al., 2004. *Gaussian 03, Revision C.02*, Wallingford, CT: Gaussian, Inc.).

2004, 789). The orientation dependence can be different for the L- and D-forms of the chiral moiety and might give rise to discrimination. The influence of various surrounding residues such as C2104, C2105, A2485, A2486, C2487, U2541, C2542, C2608, G2618, U2619, U2620, and A2637 on the rotatory path and their interaction with the sugar ring related to the A-site and P-site of the 50S ribosomal subunit of *Haloarcula marismortui* was investigated recently (Thirumoorthy and Nandi, 2008, 9187). The removal of the surrounding residues affects the nonbonded interaction with the amino acid:sugar heteropair of natural chirality. The removal of the U2620 diminished the discrimination, indicating that U2620 has favorable interaction with the L-terminal rather than the D-form of the terminal. The removal of U2620 affects the rotatory path to form a peptide bond as shown in Figure 3.14. The removal leads to loss of interaction between the sugar ring and U2620 for a certain range of orientation and hence the energy increases in the absence of U2620. The effect of the removal of A2486 on the rotatory path is presented in Figure 3.15a. The result shows loss of favorable intermolecular interaction in the range of orientation in which the rotating A-site is in proximity to the base when the latter is removed. Similar loss of intermolecular interaction is noted by removing the G2618 as shown in Figure 3.15b. On the other hand, removal of C2487, U2541, C2104, C2105, A2485, C2542, C2608, U2619, A2637, and U2541 shows no significant loss of intermolecular interaction. These results indicate that while a set of surrounding residues have favorable influence on the peptide bond formation between A- and P- terminals within PTC, a number of other residues might have less influence on the same process. However, a more detailed study using higher-level theory and basis set is necessary to understand the influence of the nucleotides on chiral discrimination in PTC.

The computational studies described earlier reveal that the molecular mechanism of the chiral discrimination involves the A- and P- amino acids, D-sugar ring, and surrounding bases of the PTC. Several factors are noted to be responsible for discrimination and explain the high level of stereospecificity of the process. The factors can be summarized as follows. The chiralities of the amino acids at A- and P- terminals are the most important. The rotatory path for the approach of D-amino acid toward the L-terminal is unfavorable due to steric hindrance between the amino acids themselves. Their nanoscale separation is important for discrimination, and at larger separation no discrimination is noted. The second factor is the restricted nature of the range of mutual orientation of the terminals during the rotatory path for the approach to form the peptide bond. This factor makes the resultant interaction profiles for L-L and D-L pairs different, with consequent discrimination. The natural chirality (D-form) of the sugar ring has a favorable influence on the long-length-scale organization of the tRNA structure and is another factor influencing discrimination. Alteration of the chirality of the sugar ring is unfavorable as it requires large structural rearrangements of tRNA. The favorable influence of the D-sugar ring is noted on the rotatory path for the process of approach to form the peptide bond. The removal of the sugar ring makes the rotatory path for approach to form the peptide bond unfavorable, which indicates that the interaction of the D-sugar with the amino acids is more favorable than other homo- or heteropair combinations of the sugar amino acid pair. The stereochemistry of the 2' center of the D-sugar has a vital influence as it catalyzes peptide bond formation by enabling proper placement of the OH group involved in the catalysis. The analysis of the transition state structure revealed that the alteration and removal of chirality of the 2' center destabilize the transition state and make the formation of the peptide bond unfavorable. Finally, the nanoscale proximity of some of the surrounding bases present in PTC with the A- and P-terminals and their restricted orientation have an influence on the discrimination. Thus, multiple factors control the discrimination in the peptide synthesis in PTC and allow accurate retention of the biological homochirality in the reaction.

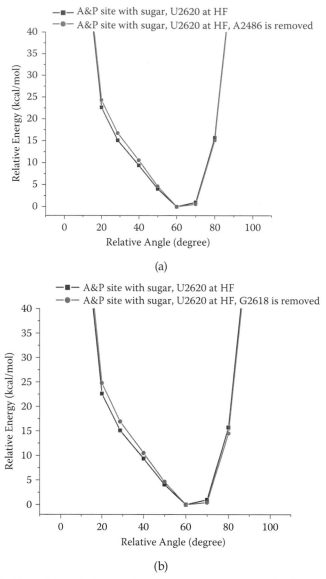

Figure 3.15 (a) Variation in interaction energy as a function of orientation of phenyl alanine at the A-terminal and phenyl alanine at the P-terminal in the presence and in the absence of A2486, respectively. (b) Variation in interaction energy as a function of orientation of phenyl alanine at the A-terminal and phenyl alanine at the P-terminal in the presence and in the absence of G2618, respectively (Thirumoorthy, K. and Nandi, N. 2008. *J. Phys. Chem. B.*, 112: 9187. Reprinted with permission from the American Chemical Society. © American Chemical Society). All calculations were performed by Gaussian 03W suite of programs (Frisch, M.J. et al., 2004. *Gaussian 03, Revision C.02*, Wallingford, CT: Gaussian, Inc.).

3.2 Chiral discrimination by telomerase

The DNA molecules that carry our genes are packed into chromosomes, the telomeres being the caps on their ends. The ends of the chromosomes are protected by the telomeres. The conventional DNA polymerases do not replicate the extreme ends of the linear DNA terminus, and therefore DNA synthesis at chromosomal ends results in progressive telomere shortening during successive rounds of cell division. Cells age if the telomeres are shortened (Blackburn, 1992, 113; Blackburn, 1991, 569; Harley et al., 1990, 458; Morin, 1989, 521). Telomerase is an enzyme (EC 2.7.7.49) that builds telomere and adds DNA sequence repeats to the 3' end of DNA strands in the telomere regions. Telomerase is classified as a reverse transcriptase. Telomerase activity is very weak in most somatic cells, but telomerase is activated in clinical cancers and cell lines significantly. It was suggested that this activation of telomerase contributes to tumor formation in vivo. Defective telomerase could give rise to the damaged cells.

Tendian and Parker studied the interaction of deoxyguanosine analogues with human telomerase (Tendian and Parker, 2000, 695). The D-carbocyclic-2'-deoxyguanosine-5'-triphosphate inhibited the telomerase activity by 50%. The corresponding L-enantiomer was far less inhibitory. Similar results for other deoxynucleoside analogues are also obtained (Pai et al., 1998, 1909). Kinetic analysis of the inhibitory effects of L-enantiomers of natural 2'-deoxyribonucleoside 5'-triphosphates (dNTP) on human telomerase enzyme was studied (Yamaguchi et al., 2000, 475). Among the four L-enantiomers of dNTPs, L-dTTP and L-dGTP inhibited telomerase activity and the other l- enantiomers showed less or no inhibitory effect. The inhibition modes of L-dTTP and L-dGTP were partially competitive and competitive with the corresponding substrate dNTP, respectively. Results from kinetic studies suggest that the active site of telomerase is not able to discriminate strictly the chiralities of the dNTPs. However, it was concluded that telomerase is more discriminatory than human immunodeficiency virus type 1 reverse transcriptase (HIV-1 RT) toward enantiomeric forms of dNTPs. Further crystal structure studies combined with computational analysis are necessary to understand discrimination in these systems.

3.3 Chiral discrimination by HIV-1 reverse transcriptase

The etiological agent of AIDS (acquired immune deficiency syndrome) is the human immunodeficiency virus (HIV). A large number of agents have been reported to inhibit the replication of HIV. Among the most potent of these are 2',3'-dideoxynucleoside analogues targeted against viral reverse transcriptase. The racemate of the β-anomer, 2′-deoxy-3′-thiacytidine (BCH 189), is a nucleoside analogue in which the ribose is replaced by a 1,3-oxathiolane ring and is active against HIV type 1 (HIV-1) in vitro. Racemic 2′-deoxy-3′-thiacytidine is a dideoxycytidine analogue having a sulfur atom in place of the 3′ carbon. The enantiomers of BCH 189 have been resolved and found to be equipotent in antiviral activity against HIV types 1 and 2 (Coates et al., 1992, 202). However, the (–)-enantiomer is considerably less cytotoxic than the (+)-enantiomer.

The inhibitory effects of 2′-deoxy-l-thymidine 5′-triphosphate (L-dTTP) on the activity of mammalian DNA polymerases α, β, and γ, *E. coli* DNA polymerase I and human immunodeficiency virus 1 (HIV-1) reverse transcriptase were examined (Yamaguchi et al., 1994, 1023). L-dTTP is the enantiomer of the natural substrate D-dTTP. When poly(rA)$_n$-oligo(dT)$_{12-18}$ was used as the template primer, L-dTTP showed a remarkable inhibitory effect on HIV-1 reverse transcriptase in a competitive fashion with respect to the substrate dTTP. In contrast, L-dTTP did not inhibit DNA polymerases α and was slightly inhibitory to DNA polymerase β. These results suggest that the nuclear DNA polymerases α and β showed high specificity for the substrate with the natural configuration of the sugar moiety, D-dTTP, exhibiting little or no ability to recognize L-dTTP, whereas HIV-1 reverse transcriptase essentially lacked the ability to differentiate between the D- and L-sugar moieties.

A lack of the enantioselectivity of human immunodeficiency virus type 1 (HIV-1) reverse transcriptase (RT) is observed. An in vitro study was carried out based on the utilization of six recombinant HIV-1 RT mutants containing single amino acid substitutions known to confer Nevirapine resistance in treated patients (Maga et al., 1999, 972). The mutants were compared on different RNA/DNA and DNA/DNA substrates to the wild-type enzyme for their sensitivity toward inhibition by the D- and L-enantiomers of 2′-deoxy- and 2′, 3′-dideoxynucleoside triphosphate analogues. The results showed that the 3′-hydroxyl group of the L-(β)-2′-deoxyribose moiety caused an unfavorable steric hindrance with critical residues in the HIV-1 RT active site. This steric barrier was increased by the Y181I mutation. Elimination of the specific hydroxyl hydroxyl group removed this hindrance and significantly improved binding. These results demonstrate the critical role of both the tyrosine 181 of RT and the 3′-position of the sugar ring in the chiral discrimination between D- and L-nucleoside triphosphates.

3.4 Chiral discrimination and nuclear DNA polymerases

Focher and coworkers studied the possible role of inhibition of viral proliferation by L-nucleosides (Focher et al., 1995, 2840). L-β-Deoxythymidine

(L-dT) is selectively phosphorylated *in vivo* to L-dTMP by herpes simplex virus type 1-TK and inhibits the proliferation of virus in infected cells. It is observed that L-dTTP not only inhibits HSV-1 DNA polymerase, in vitro, but also human DNA polymerases α, γ, δ, and ε, human immunodeficiency virus reverse transcriptase (HIV-1 RT), *E. coli*. DNA polymerase I, and Calf thymus terminal transferase. However, DNA polymerases β are resistant. The DNA polymerases β, γ, and ε are unable to utilize L-dTTP as a substrate. On the other hand, the other DNA polymerases incorporate at least one L-dTMP residue, with DNA polymerase and HIV-1 RT able to further elongate the DNA chain by catalyzing the formation of the phosphodiester bond between the incorporated L-dTMP and an incoming L-dTTP.

It has been pointed out that the viral and cellular enzymes such as herpes virus thymidine kinases, cellular deoxycytidine kinase and deoxynucleotide kinases, human immunodeficiency virus type 1 (HIV-1) reverse transcriptase, hepatitis B virus (HBV) DNA polymerase and, to a lesser extent, some cellular DNA polymerases involved in the synthesis of deoxynucleoside triphosphates and in their polymerization into DNA exhibit lack of enantioselectivity (Spadari et al., 1998, 1285).

Herpes simplex virus (HSV) thymidine kinase and cellular deoxycytidine kinase can also phosphorylate the unnatural L-β-enantiomers of D-thymidine and D-deoxycytidine, respectively, or their analogues to monophosphate. This phosphorylation represents the first and often the rate-limiting step of their activation to triphosphates. The L-triphosphates can then exert antiviral (anti-HSV, antihuman cytomegalovirus, anti-HIV-1, and anti-HBV) and anticancer effects. More studies are required to understand the effect of chirality of L-β-nucleosides toward the antiviral and anticancer activities. Such studies are relevant in solving problems arising during chemotherapy, such as metabolic inactivation, cytotoxicity, and drug-resistance where a more detailed understanding of the corresponding molecular mechanism is needed.

References

Bain, J.D., Diala, E.S., Glabe, C.G., Wacker, D.A., Lyttle, M.H., Dix, T.A., and Chamberlin, A.R. (1991). Site-specific incorporation of nonnatural residues during in vitro protein biosynthesis with semisynthetic aminoacyl-tRNAs. *Biochemistry*, 30: 5411–5421.

Berg, J.M., Tymoczko, J.L., and Stryer, L. (2003). *Biochemistry*, New York: W.H. Freeman, 814.

Bhuta, A., Quiggle, K., Ott, T., Ringer, D., and Chladek, S. (1981). Stereochemical control of ribosomal peptidyltransferase reaction: Role of amino acid side-chain orientation of acceptor substrate. *Biochemistry*, 20: 8–15.

Blackburn, E.H. (1991). Structure and function of telomeres. *Nature*, 350: 569–573.

Blackburn, E.H. (1992). Telomerases. *Annu. Rev. Biochem.*, 61: 113–129.

Calendar, R. and Berg, P. (1967). D-Tyrosyl RNA: Formation, hydrolysis and utilization for protein synthesis. *J. Mol. Biol.*, 26: 39–54.

Chamberlin, S.I. Merino, E.J., and Weeks, K.M. (2002). Catalysis of amide synthesis by RNA phosphodiester and hydroxyl groups. *Proc. Natl. Acad. Sci. (USA).*, 99: 14688–14693.

Chis, V., Pirnau, A., Jurca T., Vasilescu, M., Simon, S., Cozar, O., and David, L. (2005). Experimental and DFT study of pyrazinamide. *Chem. Phys.*, 316: 153–163.

Coates, J.A.V., Cammack, N., Jenkinson, H.J., Mutton, I.N., Pearson, B.A., Storer, R., Cameron, J.M., Penn, C.R. (1992). The separated enantiomers of 2'-Deoxy-3'-thiacytidine (BCH 189) both inhibit human immunodeficiency virus replication in vitro. *Antimicrob. Agent. Chemother.*, 36: 202–205.

Dedkova, L.M., Fahmi, N.E., Golovine, S.Y., and Hecht, S.M. (2003). Enhanced D-amino acid incorporation into protein by modified ribosomes. *J. Am. Chem. Soc.*, 125: 6616–6617.

Dorner, S., Polacek, N., Schulmeister, U., Panuschka C., and Barta, A. (2002). Molecular aspects of the ribosomal peptidyl transferase. *Biochem. Soc. Trans.*, 30: 1131–1136.

Focher, F., Maga, G., Bendiscioli, A., Capobianco, M., Colonna, F., Garbesi, A., and Spadari, S. (1995). Stereospecificity of human DNA polymerases alpha, beta, gamma, delta and epsilon, HIV reverse transcriptase, HSV-1 DNA polymerase, calf thymus terminal transferase and *Escherichia coli* DNA polymerase I in recognizing D- and L-thymidine 59-triphosphate as substrate. *Nucleic Acids Res.*, 23: 2840–2847.

Friman, K., Bohr, H., Jalkanen, K.J., and Suhai, S. (2000). Structures, vibrational absorption and vibrational circular dichroism spectra of L-alanine in aqueous solution: A density functional theory and RHF study. *Chem. Phys.*, 255: 165–194

Frisch, M.J. et al. (2004). *Gaussian 03, Revision C.02*, Wallingford, CT: Gaussian, Inc.

Gindulyte, A., Bashan, A., Agmon, I., Massa, L., Yonath, A., and Karle, J. (2006). The transition state for formation of the peptide, bond in the ribosome. *Proc. Natl. Acad. Sci (USA)*, 103: 13327–13332.

Hansen, J.L., Schmeing, M.T., Moore P.B., and Steitz, T.A. (2002). Structural insights into peptide bond formation. *Proc. Natl. Acad. Sci. (USA)*, 99: 11670–11675.

Harley, C.B., Futcher, A.B., and Greider, C.W. (1990). Telomeres shorten during ageing of human fibroblasts. *Nature*, 345: 458–460.

Harris R.J. and Pestka, S. (1977). in *Molecular Mechanisms of Protein Biosynthesis*, H. Weissbach and S. Pestka (eds), Academic Press, New York, 413–422.

Hecht, S.M. (1977). Participation of isomeric tRNA's in the partial reactions of protein biosynthesis. *Tetrahedron*, 33: 1671–1696.

Hecht, S.M. (1992). Probing the synthetic capabilities of a center of biochemical catalysis. *Acc. Chem. Res.*, 25: 545–552.

Heckler, T.G., Roesser, J.R., Xu, C., Chang, P.I., and Hecht, S. (1988). Ribosomal binding and dipeptide formation by misacylated tRNA[Phe]'s, *Biochemistry*, 27: 7254–7262.

Heckler, T.G., Zama, Y., Naka, T., and Hecht, S. (1983). Dipeptide formation with misacylated tRNA[Phe]s. *J. Biol. Chem.*, 258: 4492–4495.

Humphrey, W., Dalke, A., and Schulten, K. (1996). VMD: Visual molecular dynamics. *J. Mol. Graphics*, 14: 33–38.

Israelachvili, J.N. (1985). *Intermolecular and Surface Forces: With Applications to Colloidal and Biological Systems*, London: Academic Press.

Jalkanen, K.J., Nieminen, R.M., Friman, K., Bohr, J., Bohr, H., Wade, R.C., Tajkhorshid, E., and Suhai, S. (2001). A comparison of aqueous solvent models used in the calculation of the Raman and ROA spectra of L-alanine. *Chem. Phys.*, 265: 125–151.

Jensen, F. (1999). *Introduction to Computational Chemistry*, New York: John Wiley & Sons, 192.

Knapp-Mohammady, M., Jalkanen, K.J., Nardi, F., Wade, R.C., and Suhai, S. (1999). L-Alanyl-L-alanine in the zwitterionic state: Structures determined in the presence of explicit water molecules and with continuum models using density functional theory. *Chem. Phys.*, 240: 63–77.

Maga, G., Amacker, M., Hübscher, U., Gosselin, G., Imbach, J.-L., Mathé, C., Faraj, A., Sommadossi, J.-P., and Spadari, S. (1999). Molecular basis for the enantioselectivity of HIV-1 reverse transcriptase: Role of the 3'-hydroxyl group of the L-(β)-ribose in chiral discrimination between D- and L-enantiomers of deoxy- and dideoxy-nucleoside triphosphate analogs. *Nucleic Acids Res.*, 27: 972–978.

Maitland, G.C., Rigby, M., Smith, E.B., and Wakeham, W.A. (1981). *Intermolecular Forces: Their Origin and Determination*, Oxford: Clarendon Press.

Monro, R.E. and Marcker, K.A. (1967). Ribosome-catalysed reaction of puromycin with a formylmethionine-containing oligonucleotide. *J. Mol. Biol.*, 25: 347–350.

Monro, R.E., Cerna, J., and Marcker, K.A. (1968). Ribosome catalyzed peptidyl transfer: Substrate specificity at the P-site. *Proc. Natl. Acad. Sci. (USA)*, 61: 1042–1049.

Morin, G.B. (1989). The human telomere terminal transferase enzyme is a ribonucleoprotein that synthesizes TTAGGG repeats. *Cell*, 59: 521–529.

Nandi, N. (2004). Role of secondary level chiral structure in the process of molecular recognition of ligand: Study of model helical peptide. *J. Phys. Chem. B.*, 108: 789–797.

Nathans, D. (1964). Puromycine inhibition of protein synthesis: Incorporation of puromycin into peptide chains. *Proc. Natl. Acad. Sci. (USA)*, 51: 585–592.

Nathans, D. and Neidle, A. (1963). Structural requirements for puromycin inhibition of protein synthesis. *Nature (Lond.)*, 197: 1076–1077.

Novoa, J.J. and Sosa, C. (1995). Evaluation of the density functional approximation on the computation of hydrogen bond interactions. *J. Phys. Chem.*, 99: 15837–15845.

Pai, R.B., Pai, S.B., Kukhanova, M., Dutschman, G.E., Guo, X., and Cheng, Y. –C. (1998). Telomerase from human leukemia cells: Properties and its interaction with deoxynucleoside analogues. *Cancer Res.*, 58: 1909–1913.

Podeszwa, R. and Szalewicz, K. (2005). Accurate interaction energies for argon, krypton, and benzene dimers from perturbation theory based on the Kohn–Sham model. *Chem. Phys. Lett.*, 412: 488–493.

Quiggle, K., Kumar, G., Ott, T.W., Ryu, E.K., and Chladek, S. (1981). Donor site of ribosomal peptidyltransferase: Investigation of substrate specificity using 2' (3')-O-(N-Acylaminoacyl) dinucleoside phosphates as models of the 3' terminus of N-acylaminoacyl transfer ribonucleic acid. *Biochemistry*, 20: 3480–3485.

Rodnina, M.V., Beringer, M., and Bieling, P. (2005). Ten remarks on peptide bond formation on the ribosome. *Biochem. Soc. Trans.*, 33: 493–498.

Roesser, J.R., Xu, C., Payne, R.C. Surratt, C.K., and Hecht, S. (1989). Preparation of misacylated aminoacyl-*t*RNA^Phe's useful as probes of the ribosomal acceptor site. *Biochemistry*, 28: 5185–5195.

Schmeing, T.M., Huang, K.S., Kitchen, D.E., Strobel, S.A., and Steitz, T.A. (2005). Structural insights into the roles of water and the 2′-hydroxyl of the P site tRNA in the peptidyl transferase reaction. *Mol. Cell.*, 20: 437–448.

Schweet, R.S. and Allen, E. (1958). Purification and properties of tyrosine-activating enzyme of hog pancreas. *J. Biol. Chem.*, 233: 1104–1108.

Sharma, P.K., Xiang, Y., Kato, M., and Warshel, A. (2005). What are the roles of substrate-assisted catalysis and proximity effects in peptide bond formation by the ribosome? *Biochemistry*, 44: 11307–11310.

Sievers, A., Beringer, M., Rodnina, M.V., and Wolfenden, R. (2004). The ribosome as an entropy trap. *Proc. Natl. Acad. Sci. (USA)*, 101: 7897–7901.

Spadari S., Maga G., Verri, A., and Focher, F. (1998). Molecular basis for the antiviral and anticancer activities of unnatural L-beta-nucleosides. *Expert Opin. Investig. Drugs.*, 8: 1285–1300.

Sprinzl, M. and Cramer, F. (1979). The -C-C-A end of tRNA and its role in protein biosynthesis. *Prog. Nucleic Acid Res. Mol. Biol.*, 22: 1–69.

Starck, S.R., Qi, X., Olsen, B.N., and Roberts, R.W. (2003). The puromycin route to assess stereo- and regiochemical constraints on peptide bond formation in eukaryotic ribosomes. *J. Am. Chem. Soc.*, 125: 8090–8091.

Tajkhorshid, E., Jalkanen, K.J., and Suhai, S. (1998). Structure and vibrational spectra of the zwitterion L-alanine in the presence of explicit water molecules: A density functional analysis. *J. Phys. Chem.*, 102: 5899–5913.

Tendian, S.W. and Parker, W.B. (2000). Interaction of deoxyguanosine nucleotide analogs with human telomerase. *Mol. Pharmacol.*, 57: 695–699.

Thirumoorthy, K. and Nandi, N. (2006). Comparison of the intermolecular energy surfaces of amino acids: Orientation-dependent chiral discrimination. *J. Phys. Chem. B.*, 110: 8840–8849.

Thirumoorthy, K. and Nandi, N. (2007a). Homochiral preference in peptide synthesis in ribosome: Role of amino terminal, peptidyl terminal and U2620. *J. Phys. Chem. B*, 111: 9999–10004.

Thirumoorthy, K. and Nandi, N. (2007b). Water catalyzed peptide bond formation in L-alanine dipeptide: The role of weak hydrogen bonding. *J. Mol. Str. (Theo Chem)*, 818: 107–118.

Thirumoorthy, K. and Nandi, N. (2008). Role of chirality of the sugar ring in the ribosomal peptide synthesis. *J. Phys. Chem. B.*, 112: 9187–9195.

Trobro, S. and Åqvist, J. (2005). Mechanism of peptide bond synthesis on the ribosome. *Proc. Natl. Acad. Sci. (USA)*, 102: 12395–12400.

van Duijneveldt, F.B.V., van Duijneveldt-van de Rijdt, J.G.C.M., and van Lenthe, J.H. (1994). State of the art in counterpoise theory. *Chem. Rev.*, 94: 1873–1885.

Weinger, J.S. and Strobel, S. (2007). Exploring the mechanism of protein synthesis with modified substrates and novel intermediate mimics. *Blood Cells Mol. Dis.*, 38: 110–116.

Weinger, J.S., Parnell, K.M., Dorner, S., Green, R., and Strobel, S. (2004). Substrate assisted catalysis of peptide bond formation by the ribosome. *Nature Struc. Mol. Biol.*, 11: 1101–1106.

Williams, H.L. and Chabalowski, C.F. (2001). Using Kohn-Sham orbitals in symmetry-adapted perturbation theory to investigate intermolecular interactions. *J. Phys. Chem. A.*, 105: 646–659.

Yamaguchi T., Iwanami N., Shudo, K., and Saneyoshi, M. (1994). Chiral discrimination of enantiomeric 2'-deoxythymidine 5'-triphosphate by HIV-1 reverse transcriptase and eukaryotic DNA polymerases. *Biochem. Biophys. Res. Commun.*, 200: 1023–1027.

Yamaguchi, T., Yamada, R., Tomikawa, A., Shudo, K., Saito, M. Ishikawa, F., and Saneyoshi, M. (2000). Recognition of 2'-deoxy-L-ribonucleoside 5'-triphosphates by human telomerase. *Biochem. Biophys. Res. Commun.*, 279: 475–481.

Yamane, T., Miller, L., and Hopfield, J.J. (1981). Discrimination between D- and L-tyrosyl transfer ribonucleic acids in peptide chain elongation. *Biochemistry*, 20: 7059–7064.

Youngman, E.M., Brunelle, J.L., Kochaniak, A.B., and Green, R. (2004). The active site of the ribosome is composed of two layers of conserved nucleotides with distinct roles in peptide bond formation and peptide release. *Cell*, 117: 589–599.

Zarivach, R., Bashan, A., Berisio, R., Harms, J., Auerbach, T., Schluenzen, F., Bartels, H., Baram, D., Pyetan, E., Sittner, A., Amit, M., Hansen, H.A.S., Kessler, M., Liebe, C., Wolff, A., Agmon, I., and Yonath, A. (2004). Functional aspects of ribosomal architecture: Symmetry, chirality and regulation. *J. Phys. Org. Chem.*, 17: 901–912.

Zhang, S.G. and Yang, P. (2005). Structures and properties of cytosine–BX_3 (X=F, Cl) complexes: An investigation with DFT and MP2 methods. *J. Mol. Struc. (Theo Chem)*, 757: 77–86.

chapter four

Influence of chirality on the hydrolysis reactions within the active site of hydrolases

Hydrolases have a superior ability to discriminate molecular chirality compared to many other enzymes used for enantioselective processes. They are highly promising as industrially viable processes based on hydrolase-catalyzed resolution (Bornsheuer and Kazlauskas, 1999). The hydrolytic enzymes, such as lipases, proteases, and esterases, are highly useful as chiral catalysts for enantioselective synthesis, and several case studies are reported (Chikusa et al., 2003, 289). However, in many cases, the development of the enantioselection is based on trial and error, and the details of the active site of the hydrolase and its influence on the course of the reaction are yet to be understood. In this chapter, a few examples of chiral discrimination by hydrolases are discussed where the molecular origin of the chiral preference is investigated.

4.1 Chiral discrimination by epoxide hydrolases

Epoxide hydrolases (EHs) act upon epoxides (also termed *alkylene oxides* or *oxiranes*) and convert them to their corresponding diols using a water molecule (Archer, 1997, 15617; Archelas and Furstoss, 1998, 108). These enzymes are important for the detoxification of xenobiotic compounds, which have harmful effects (mutagenic and carcinogenic initiators). Distinct subtypes of epoxide hydrolase in higher organisms include the plant-soluble epoxide hydrolase, the mammalian soluble epoxide hydrolase, the hepoxilin hydrolase, leukotriene A4 hydrolase, the microsomal epoxide hydrolase, and the insect juvenile hormone epoxide hydrolase (Newman et al., 2005, 1). Mammalian EHs detoxify epoxides in the liver. These enzymes have enormous potential in biocatalysis. They can resolve racemic mixtures of epoxides with high enantiomeric excesses. The active sites of the EHs are studied in detail using computational methods with the aim of addressing the mechanistic issues and regioselectivity (Schiøtt and Bruice, 2002, 14558; Lau et al., 2001, 3350). The active site of the enzyme is located within two domains and is characterized by the presence of conserved Asp-His-Asp residues as well as two Tyr residues. The

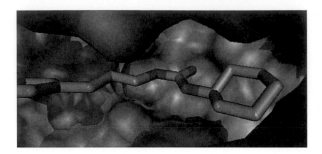

Figure 4.1 **See color insert.** Active site of liver cytosolic epoxide hydrolase (1CR6. PDB). The image is generated using VMD (Humphrey et al., 1996: 33).

topological features of the active sites from different species are remarkably similar. Various ground-state structures close to the transition state (termed *near attack conformations*) are explored in detail. It is indicated that more near attack conformations in an enzyme–substrate complex indicate higher enzyme efficacy. Since such conformations are dependent on the active site structure, it is quite possible that the chirality of the active site residues influences the discrimination, as exhibited remarkably by hydrolases. The active site structure of liver cytosolic epoxide hydrolase (Argiriadi et al., 1999, 10637) is shown in Figure 4.1.

Kinetic studies of the enantioselective hydrolysis of styrene oxide by Epicholorohydrin epoxide hydrolase (EchA) from *Agrobacterium radiobacter*, AD1, was carried out (Rink and Janssen, 1998, 18119). EchA could discriminate between the enantiomers of styrene oxide and substituted variants, leading to the production of enantiomerically pure epoxides. The interesting feature of the kinetics of conversion of styrene oxide is that the R-enantiomer is completely hydrolyzed before the S-enantiomer, and the latter is subsequently converted at a much higher rate.

This sequential hydrolysis of a racemate in which the second enantiomer is hydrolyzed faster was also noted for the hydrolysis of styrene oxide and *tert*-butyloxirane by microsomal epoxide hydrolase (Watabe et al., 1981, 1695; Wistuba and Schurig, 1992, 178; Wistuba et al., 1992, 185). The sequential hydrolysis was studied in detail by kinetic measurements. It was observed that the fluorescence trace from the single-turnover reaction with S-styrene oxide is distinctly different from the trace for R-styrene oxide, since no transient increase in the fluorescence signal is observed with the S-enantiomer. The rapid-quench data of S-styrene oxide showed the formation of covalent intermediate. A multistep mechanism has been proposed for R-styrene oxide. First, the formation of a Michaelis complex that is in rapid equilibrium with free enzyme and substrate is followed by rapid and reversible alkylation of the enzyme. The alkylated enzyme undergoes isomerization before the hydrolysis of the covalent intermediate takes place. On the other hand, a three-step mechanism is suggested for the conversion of S-styrene oxide, involving reversible and rapid formation of an ester intermediate from a Michaelis complex followed by its subsequent rate-limiting hydrolysis step.

It has been proposed that during the reaction, Asp107 in the active site makes a nucleophilic attack on the primary carbon atom of the epoxide ring and forms a covalently bound ester intermediate. The Asp246-His275 pair, supported by Asp131, activates a water molecule that hydrolyzes the ester bond at the carbonyl group of Asp107 followed by the release of product. The more pronounced enantioselectivity in the alkylation rate compared to the rate-limiting hydrolysis step might be related to the chiral discrimination of the former step, which depends on the active site structure. The active site structure of the EHs from *Agrobacterium radiobacter*, AD1 was analyzed (Nardini et al., 1999, 14579). The reaction center includes residues such as multiple Asp, His, Trp, Phe, and Tyr. Although it is expected that the presence of these chiral moieties have an influence on the different enantiomers of styrene oxide, more molecular studies are necessary to understand this remarkable discrimination phenomenon.

Despite the similarity of the mechanisms of EchA and rat microsomal epoxide hydrolase enzyme actions, the origin of the enantioselectivity has been suggested to be different in the two cases (Tzeng et al., 1996, 9436). The enantioselectivity of the enzyme has been observed in both the alkylation and hydrolysis steps. However, this selectivity is most significantly manifested in the hydrolysis of the intermediate. The regiospecific lty for the reaction is the same because the carboxylate group attacks the primary oxirane carbon in the cases of both enantiomers. The active-site nucleophile has been suggested as Asp226 for microsomal epoxide hydrolase. However, it is yet to address the molecular mechanism of the differences in the chiral discrimination features observed in the cases of rat microsomal epoxide hydrolase and those observed in Epicholorohydrin epoxide hydrolase from *Agrobacterium radiobacter*.

Conversion of R/S-styrene oxide by EchA has the special feature that the R-enantiomer is converted and hydrolyzed and subsequently the S-enantiomer starts hydrolyzing. During the hydrolysis of the R-enantiomer, the conversion of the S-enantiomer is inhibited. The latter is subsequently hydrolyzed. More interestingly, once the S-enantiomer starts converting, it hydrolyzes at a much faster rate (Lutje et al., 1998, 459). This feature is expected to arise from the capacity of the active site to discriminate between the interactions with the R- and S-enantiomers. The result suggests that the R-enantiomer effectively binds to the active site, and once its hydrolysis is completed, the S-enantiomer binds to the active site. However, the reaction mechanism of the S-enantiomer within the active site is such that the overall kinetics is faster. For R-styrene oxide, a four-step mechanism was proposed to describe the data. It involved the formation of a Michaelis complex that is in rapid equilibrium with free enzyme and substrate, followed by rapid and reversible alkylation of the enzyme. A unimolecular isomerization of the alkylated enzyme precedes the hydrolysis of the covalent intermediate. The conversion of S-styrene oxide could be described by a three-step mechanism, which also involved reversible and rapid formation of an ester intermediate from a Michaelis complex and its subsequent slow hydrolysis as the rate-limiting step. The unimolecular isomerization step has not been observed for rat microsomal epoxide hydrolase, for which a kinetic mechanism was recently established (Tzeng et al., 1996, 9436). For both enantiomers of styrene oxide, the K_m value was much lower than the substrate binding constant K_s due to extensive accumulation of the covalent intermediate. The enantioselectivity was more pronounced in the alkylation rates than in the rate-limiting hydrolysis steps. The combined reaction schemes for R- and S-styrene oxide gave an accurate description of the epoxide hydrolase catalyzed kinetic resolution of racemic styrene oxide.

Detailed studies using hybrid density functional theory (B3LYP) of mutation of the active site have been carried out to understand the molecular basis of the enzymatic action of the human soluble epoxide hydrolase. It has been observed that two conserved active site tyrosines lower the activation barrier for the alkylation step. A histidine has been suggested to have a role for the alkylation half-reaction (Hopmann and Himo, 2006, 21299). Since the Tyr, His, and other residues might be involved in different steps of the reaction, it is expected that the reaction would exhibit remarkable chiral specification.

4.2 *Chiral discrimination by lipases*

Lipases are important for intestinal lipolysis of triacylglycerols. Lipases (EC 3.1.1.3.) hydrolyze ester bonds present in neutral lipids. These esterases have the remarkable property that they exhibit significant activity in the presence of lipid–water interfaces, which is known as interfacial activation. Lipases can be used not only in water, but also in water–organic solvent mixtures or even in pure anhydrous organic solvents when a sufficient amount of water is available to maintain the corresponding catalytic conformation. Lipases have potential in asymmetric catalysis as homochiral compounds can be obtained in principle as the product of biocatalysis. Digestive lipases are very important for intestinal absorption of dietary triacylglycerols in higher animals. Polar fatty acids and monoacylglycerols are digested by the enzyme action of preduodenal and pancreatic lipases. A number of examples of the enantioselectivity of lipases are reported in the literature (Gargouri et al., 1997, 6). In the following text, we mention some examples where the role of the active site of the lipases on the chiral selectivity of the enzyme is indicated.

The interaction of lipases with substrates has been proposed to involve two distinct steps: adsorption to the lipid–water interface and subsequent binding (and hydrolysis) of a single substrate molecule in the active site. Most lipases contain buried active sites with a lid or cap covering the active site, which is composed of the catalytic triad Asp (or Glu)-His-Ser having a catalytic role. The Ser is located at a hairpin turn between an alpha-helix and a beta strand, and the Asp (or Glu) and His reside on the other side. This lid covers the active site when the lipase is present in solution. However, the flap opens in the presence of lipid or organic solvent, and the hydrolysis rate increases as the active site is exposed. The folding patterns of different lipases such as *Rhizomucor miehei* lipase, human pancreatic lipase, *Geotrichum candidum* lipase, *Candida rugosa* lipase and *Pseudomonas glumae* lipase, cutinase, and acetylcholinesterase bear close resemblance in their active site structure in general and the location of the Asp (or Glu)-His-Ser catalytic triad despite the difference in the primary sequence (Cygler et al., 1994, 3180).

Mannesse and co-workers studied the inhibition of cutinase from *Fusarium solani pisi* and *Staphylococcus hyicus* lipase by 1,2-dioctylcarbamoylglycero-3-O-p-nitrophenyl methyl/octyl-phosphonates (Mannesse et al., 1995, 56). Both lipases exhibit high chiral selectivity of these compounds in glycerol and phosphorous. Rapid inactivation occurred when the glycerol moiety had R-configuration. The inhibitors with S-configuration react four- to tenfold more slowly. The isomer with the p-nitrophenyl octylphosphonate attached to the secondary hydroxyl group of glycerol inhibited the lipases far less. These enzymes preferentially release the fatty acid at sn-3 of triacylglycerols. This indicates the positional as well as stereoselectivity of the enzyme. For the methylphosphonates, the difference in reactivity between the faster and slower reacting isomers is about 250-fold. For octylphosphonates, the difference is 60-fold. This high level of discrimination indicates the higher selectivity of the enzyme for the phosphorus chirality. It has been indicated that these phosphonates can be regarded as true active-site-directed inhibitors.

Both lipases recognize only one asymmetric configuration around phosphorus, or at least possess a strong preference for either the R_p or S_p isomer of the inhibitor. These enzymes differentiate between the R_p/S_p isomers of the large O-diacylglycerol-substituted phosphonates, whereas they do not recognize the chirality at phosphorus of the smaller O-methylphosphonates. The differences in behavior toward the two types of phosphonates can be understood in terms of the proposed presence of two binding pockets for the acyl chains of the substrate in lipases. It has been suggested that two acyl ester chains bind to hydrophobic protein surfaces. The third acyl ester chain points into solution and might be still partially embedded in the substrate–water interface. In such a case, the single alkyl chains of both enantiomers might occupy two different binding sites. However, the position of the phosphorus atom near the nucleophilic active site serine is almost identical in both configurations. In such a situation, recognition of the chirality at phosphorus by the enzyme is improbable, and very similar inactivation rates by the R_p/S_p isomers can be expected. However, when the small ester group in the phosphonate is replaced by the bulky enantiomeric 1,2 diacylglycerol analogues, the freedom of both alkyl chains within the two hydrophobic grooves on the enzyme is restricted. In this case, the placement of the phosphorus atom near the active site serine become highly specific, which in turn enables precise binding of the alkyl chains of diacylglycerol inhibitors.

A detailed overview of the discrimination of enantiomers from racemic carboxylic esters by lipases from fungi such as *Rhizomucor miehei* lipase is reviewed in the literature (Alcántara et al., 1998, 169). Both hydrolysis and acyl transfer mechanisms are discussed. In the case of hydrolysis of racemic esters with the stereocenter at the acid moiety, the carboxylic group of the ester is attached to a secondary carbon and one aromatic substituent is required in the typical substrate recognition structure. In the case of chiral racemic acids also, the configuration of the preferred enantiomer depends on the presence of an aromatic group in the asymmetric carbon atom. 2-aryl propionic acid conformers are studied by simulation and molecular mechanics program to understand the recognition process. Recognitions of the S-enantiomers in this group are very similar, except for S-ketoprofen (its benzoyl group is located far away from the aromatic area of the S-conformers of the other compounds). On the other hand, the benzoyl group of R-ketoprofen is located closer to the other aromatic rings, though hydrogen and methyl are on opposite sides. However, such hypotheses are yet to be further substantiated using analysis of interactions in the respective molecular structure and active site residues. Hence, possible discriminations due to the variations in interactions are yet to be understood.

Lang and Dijkstra determined the x-ray structures of *Pseudomonas cepacia* lipase in complex with two enantiopure triglyceride analogues that closely mimic natural substrates (Lang and Dijkstra, 1998, 115). This allowed an unambiguous view of how the two wings of the boomerang-shaped active site accommodate the acyl and alcohol parts of the triglyceride. The binding groove for the hydrophobic sn-3 fatty acid chain is large and hydrophobic. The cleft for the alcohol moiety is divided into two parts. The first one tightly binds the sn-2 acyl chain, which may undergo hydrophilic and hydrophobic interactions with the active site residues, and the other weakly binds the sn-1 fatty acid. This indicates that the confinement of different chains of the enantiomeric species into specific regions of the actives site is a key factor of the chiral discrimination capacity of the lipase. The enantioselectivity of *Pseudomonas cepacia* lipase seems therefore to be predominantly determined by the size and interactions of the sn-2 chain and by the size of the sn-3 chain.

In a series of detailed studies, Grochulski and co-workers determined the structure of *Candida rugosa* lipase in the unliganded state in both open and closed form as well as complexes with inhibitor (Grochulski et al., 1993, 12843; Grochulski et al., 1994, 82). The lipase is in the inactive state when the conformation of the solvent accessible active site is shielded from the solvent by the flap that is a part of the polypeptide chain. The flap is flexible, with hinge points at Glu66 and Pro92, accompanied by a cis-trans isomerization of the latter residue. When the flap moves, the active site turns into the open state with a large increase of the hydrophobic surface in the vicinity of the active site. While some of the secondary structure of the flap refolds during the flap movement, the backbone NH groups forming the putative oxyanion hole do not change position during this reorganization. The structures of *Candida rugosa* lipase–inhibitor complexes were also studied, which showed that the fatty acyl chain is bound in a narrow, hydrophobic tunnel. Modeling of triglyceride binding suggested that the lipid must take up a conformation similar to the shape of a tuning fork in the bound state (Grochulski et al., 1994, 3494).

The foregoing studies provided a molecular basis for the highly efficient chiral discrimination exhibited by the lipases within their active site. Cygler and co-workers showed that the *Candida rugosa* lipase can catalyze the preferential hydrolysis of (1R,2S,5R) menthyl pentanoate to (1R,2S,5R) methol over (1S,2R,5S) menthyl pentanoate. The molecular basis of the chiral preference of lipases was studied in detail. The observed enantio-preference was attempted to be predicted empirically (Cygler et al., 1994, 3180). The empirical rule is successful in predicting which enantiomer of a secondary alcohol can react faster with *Candida rugosa* lipase and other selected lipases. The substrate is the ester of an alcohol in the hydrolysis reaction and an alcohol in the transesterification reaction. In the case of alcohol as a substrate, the favored enantiomer should have the hydroxyl group projected out the plane of the paper, the large group located on the right side on the plane of the paper and the medium-sized group located on the left-hand side of the plane of the paper. This empirical rule is supported by the fact that while the lipases resolve the secondary alcohols with unequal-sized substituents efficiently, the alcohols with substituents resolve poorly.

A molecular-level understanding of the empirical rule is provided from the x-ray crystal structures of covalent complexes of *Candida rugosa* lipase with transition-state analogues for the hydrolysis of menthyl esters (Cygler et al., 1994, 3180). The overall shape of the active site cavity located at the C-terminal end of the beta strands is oval-shaped. The large substituent (indicated by "Large" in Figure 4.2) is located in the hydrophobic region lined with phenyl rings, and the medium-sized substituent

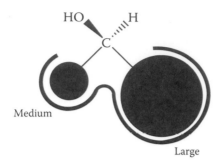

Figure 4.2 Schematic representation of the recognition of enantiomers of secondary alcohols by *Candida rugosa* lipase (Cygler, M. et al., 1994. *J. Am. Chem. Soc.*, 116: 3180–3186). The hydroxyl group is pointed toward the observer, "Medium" represents a medium group, and "Large" indicates a relatively large group.

(indicated by "Medium" in Figure 4.2) is placed on the floor of the active site. The possible catalytic role of amino acid residues in the active site is considered. It is observed that the hydrophobic pocket is flexible enough to accommodate different substituents. Analysis of the orientation of the imidazole ring of the His of the catalytic triad suggests that the interactions between the menthyl ring of the slow-reacting enantiomer and the histidine of the catalytic triad disrupt the hydrogen bond between Nε2 of the imidazole ring, and the menthol oxygen atom could be responsible for the slower reaction of the 1S-enantiomer of menthol. The enantiopreference of *Candida rugosa* lipase toward secondary alcohols is controlled by the same loops that assemble the catalytic machinery. The common orientation of these loops among many lipases and esterases accounts for their common enantiopreference toward secondary alcohols. Despite the apparent success of the size-based understanding of the mechanism of the chiral discrimination, it would be helpful to analyze the intermolecular interaction in the active site for rational enantioselective syntheses with lipases.

The structure of the active site indicates that the confinement of the lipid chains in a suitable orientation for enzymatic reaction within the crevices of the active site is an important factor for the highly efficient discrimination processes exhibited by lipases. Computational studies have indicated that the confinement of a chiral molecule near neighboring chiral entities is important for an efficient discrimination process (Nandi, 2004, 789; Nandi, 2009, 111). The mechanism of discrimination exhibited by lipases as understood from crystallographic analysis corroborates this view. However, it must be pointed out that the discrimination exhibited by lipases is not entirely size based. Favorable interaction between the different groups of the enantiomer with the residues forming the crevices of the active site must be a determining factor.

References

Alcántara, A.R., de Fuentes, I.E., and Sinisterra, J.V. (1998). *Rhizomucor miehei* lipase as the catalyst in the resolution of chiral compounds: An overview. *Chem. Phys. Lipids,* 93: 169–184.

Archelas, A. and Furstoss, R. (1998). Epoxide hydrolases: New tools for the synthesis of fine organic chemicals. *Trends Biotechnol.,* 16: 108–116.

Archer, I.V.J. (1997). Epoxide hydrolases as asymmetric catalysts. *Tetrahedron,* 53: 15617–15662.

Argiriadi, M.A., Morisseau, C., Hammock, B.D., and Christianson, D.W. (1999). Detoxification of environmental mutagens and carcinogens: Structure, mechanism and evolution of liver epoxide hydrolase. *Proc. Natl. Acad. Sci. (USA),* 96: 10637–10642.

Bornsheuer, U.T. and Kazlauskas, R.J. (1999). *Hydrolases in Organic Synthesis: Regio- and Stereoselective Biotransformations,* Wiley-VCH: Weinheim.

Chikusa, Y., Hirayama, Y., Ikunaka, M., Inoue, T., Kamiyama, S., Moriwaki, M., Nishimoto, Y., Nomoto, F., Ogawa, K., Ohno, T., Otsuka, K., Sakota, A.K., Shirasaka, N., Uzura, A., and Uzura, K. (2003). There's no industrial biocatalyst like hydrolase: Development of scalable enantioselective processes using hydrolytic enzymes. *Org. Process Res. Development,* 7: 289–296.

Cygler, M., Grochulski, P., Kazlauskas, R.J., Schrag, J.D., Bouthillier, F., Rubin, B., Serreqi, A.N., and Gupta, A.K. (1994). A structural basis for the chiral preferences of lipases. *J. Am. Chem. Soc.,* 116: 3180–3186.

Gargouri, Y., Ransac, S. and Verger, R. (1997). Covalent inhibition of digestive lipases: An in vitro study. *Biochim. Biophys. Acta.,* 1344: 6–37.

Grochulski, P., Bouthillier, F., Kazlauskas, R.J., Serreqi, A.N., Schrag, J.D., Ziomek, E., and Cygler, M. (1994). Analogs of reaction intermediates identify a unique substrate binding site in *Candida rugosa* Lipase. *Biochemistry,* 33: 3494–3500.

Grochulski, P., Li, Y., Schrag, J.D., Bouthillier, F., Smith, P., Harrison, D., Rubin, B., and Cygler, M. (1993). Insights into interfacial activation from an open structure of *Candida rugosa* lipase. *J. Biol. Chem.,* 268: 12843–12847.

Grochulski, P., Li, Y., Schrag, J.D. and Cygler, M. (1994). Two conformational states of *Candida rugosa* lipase. *Protein Sci.,* 3: 82–91.

Hopmann, K.H. and Himo, F. (2006). Insights into the reaction mechanism of soluble epoxide hydrolase from theoretical active site mutants. *J. Phys. Chem. B.,* 110: 21299–21310.

Humphrey, W., Dalke, A., and Schulten, K. (1996). VMD: Visual molecular dynamics. *J. Mol. Graphics,* 14: 33–38.

Lang, D.A. and Dijkstra, B.W. (1998). Structural investigations of the regio- and enantioselectivity of lipases. *Chem. Phys. Lipids,* 93: 115–122.

Lau, E.Y., Newby, Z.E., and Bruice, T.C. (2001). A theoretical examination of the acid-catalyzed and noncatalyzed ring-opening reaction of an oxirane by nucleophilic addition of acetate. Implications to epoxide hydrolases. *J. Am. Chem. Soc.,* 123: 3350–3357.

Lutje Spelberg, J.H., Rink, R., Kellogg, R.M., and Janssen, D.B. (1998). Enantioselectivity of a recombinant epoxide hydrolase from *Agrobacterium radiobacter. Tetrahedron: Asymmetry,* 9: 459–482.

Mannesse, M.L.M., Boots, J-W.P., Dijkman, R., Slotboom, A.J., van der Hijden, H.T.W.M., Egmond, M.R., Verheij, H.M., and de Haas, G.H. (1995). Phosphonate analogues of triacylglycerols are potent inhibitors of lipase. *Biochim. Biophys. Acta*, 1259: 56–64.

Nandi, N. (2004). Role of secondary level chiral structure in the process of molecular recognition of ligand: Study of model helical peptide. *J. Phys. Chem. B.*, 108: 789–797.

Nandi, N. (2009). Chiral discrimination in the confined environment of biological nanospace: Reactions and interactions involving amino acids and peptides. *Int. Rev. Phys. Chem.*, 28: 111–167.

Nardini, M., Ridder, I.S., Rozeboom, H.J., Kalk, K.H., Rink, R., Janssen, D.B., and Dijkstra, B.W. (1999). The x-ray structure of epoxide hydrolase from *Agrobacterium radiobacter* AD1: An enzyme to detoxify harmful epoxides. *J. Biol. Chem.*, 274: 14579–14586.

Newman, J.W., Morisseau, C., and Hammock, B.D. (2005). Epoxide hydrolases: Their roles and interactions with lipid metabolism. *Prog Lipid Res.* 44: 1–51.

Rink, R. and Janssen, D.B. (1998). Kinetic mechanism of the enantioselective conversion of styrene oxide by epoxide hydrolase from *Agrobacterium radiobacter* AD1. *Biochemistry*, 37: 18119–18127.

Schiøtt, B. and Bruice, T.C. (2002). Reaction mechanism of soluble epoxide hydrolase: Insights from molecular dynamics simulations. *J. Am. Chem. Soc.*, 124: 14558–14570.

Tzeng, H.-F., Laughlin, L.T., Lin, S., and Armstrong, R.M. (1996). The catalytic mechanism of microsomal epoxide hydrolase involves reversible formation and rate-limiting hydrolysis of the alkyl-enzyme intermediate. *J. Am. Chem. Soc.*, 118: 9436–9437.

Watabe, T., Ozawe, N., and Hiratsuka, A. (1981). Stereochemistry in the oxidative metabolism of styrene by hepatic microsomes. *Biochem. Pharmacol.*, 30: 1695–1698.

Wistuba, D. and Schurig, V. (1992). Enantio- and regioselectivity in the epoxide hydrolase catalyzed ring opening of simple aliphatic oxiranes: Part I: Monoalkyl substituted oxiranes. *Chirality*, 4: 178–184.

Wistuba, D., Träger, O. and Schurig, V. (1992). Enantio- and regioselectivity in the epoxide hydrolase catalyzed ring opening of simple aliphatic oxiranes: Part II: Dialkyl and trialkylsubstituted oxiranes. *Chirality*, 4: 185–192.

chapter five

Influence of chirality on the reactions in the active site of lyases

Lyases are a class of enzymes that catalyze the breaking of various chemical bonds by molecular mechanisms that are different from the process of hydrolysis and oxidation, often forming a new double bond or a new ring structure. The common names of lyases are decarboxylase, dehydratase, or aldolase. The term *synthase* is used for the reverse reaction. The breaking of the adenosine triphosphate (ATP) into adenosine monophosphate (AMP) and pyrophosphate (PPi) is an example of a reaction carried out by the lyases. This reaction is an example that shows the feature that if a lyase requires one substrate in the forward direction of the reaction, then it requires two substrates in the reverse direction.

5.1 Hydroxynitrile lyases: interaction with chiral substrates

Hydroxynitrilase or α-hydroxynitrile lyase (HNL) enzymes are available from different natural sources and are useful for industrial-scale biocatalysis. They are isolated from plants such as almonds, millet, rubber tree, flax, and cassava. These enzymes release HCN (cyanogenesis). Cyanogenesis occurs in injured tissues of plant species, including crop plants to protect against predation or fungal attack. Stored cyanoglycosides produce free sugar and α-hydroxynitrile (cyanohydrin) initiated by a β-glycosidase. Chiral cyanohydrins are important synthetic intermediates for the production of a wide range of pharmaceuticals and agrochemicals (Griengl et al., 2000, 252). The cyanohydrin is unstable and further decomposes into HCN and the corresponding aldehyde or ketone. The reaction is spontaneous at high pH, but requires the catalysis by the α-hydroxynitrile lyase (HNL) enzyme at low pH.

It is possible to isolate R-oxynitrilase in larger amounts from different natural sources, for example, bitter and sweet almonds. However, S-oxynitrilases cannot be obtained in large quantities from natural sources, as their concentrations in cells are small. Recombinant microorganisms are useful in producing these enzymes in larger quantities such

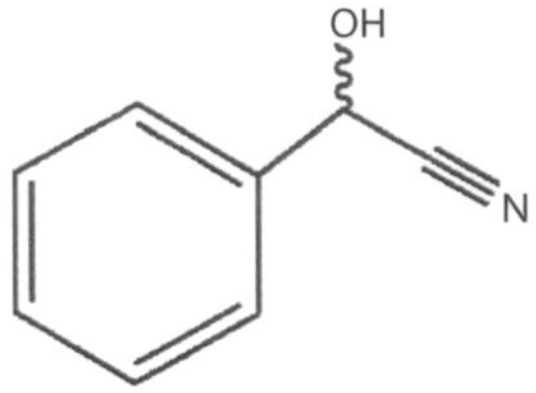

Figure 5.1 Chemical structure of mandelonitrile (2-hydroxy 2-phenyl acetonitrile).

as expression systems for S-oxynitrilase from the rubber tree (*H. brasiliensis*) or the recombinant *E. coli* system that can produce S-oxynitrilase from *Manihot esculenta* (Daußmann et al., 2006, 125).

As early as 1908, Rosenthaler reported the preparation of R-mandelonitrile from benzaldehyde and hydrogen cyanide using emulsin as catalyst (Rosenthaler, 1908, 238). Mandelonitrile is a compound of the cyanohydrin class (Figure 5.1) and can be broken down into cyanide and benzaldehyde by lyase. Emulsin is a mixture of enzymes in bitter almond that hydrolyze the glucoside amygdalin to benzaldehyde, glucose, and cyanide. This study is one of the earliest descriptions of an asymmetric biocatalysis reaction (asymmetric addition of HCN to benzaldehyde by R-oxynitrilases from almond extract). Chiral α-hydroxy acids can be synthesized via asymmetric addition of hydrocyanic acid to aldehydes catalyzed by HNL enzyme such as S-oxynitrilase from *Manihot esculenta*. In synthetic applications, oxynitrilases are used for the asymmetric addition of HCN to carbonyl groups. Asymmetric HCN addition to aliphatic and α,β-unsaturated aldehydes can be achieved often with enantiomeric excess of more than 90%. Lower enantioselectivities are observed for ketones. The cyanohydrin formed bears an S-configuration. Organic solvents are effective in producing enantiomeric excess because the racemic addition of HCN to the carbonyl group is suppressed. The active site of the enzyme contains a catalytic triad consisting of serine, histidine, and asparagine. In contrast to all other enzymes having a catalytic triad, HNLs catalyze no net hydrolytic reactions. Mutant structures are characterized by improved features with respect to enantioselectivity and other features such as substrate range and kinetic data. Hence, different mutagenesis techniques for the development of new oxynitrilases are expected to be useful (Daußmann et al., 2006, 125).

The hydroxynitrile lyase from the tropical rubber tree *Hevea brasiliensis* (HbHNL) is an important biocatalyst for stereospecific synthesis of α-hydroxynitriles from aldehydes and ketones. This reaction is an example of an enzyme-mediated C-C coupling reaction that has industrial relevance. The crystal structure of recombinant HNL from *Hevea brasiliensis* was determined, and the mechanism of the cyanogenic enzyme can be understood (Wagner et al., 1996, 811). The active site is deeply buried inside the protein and is linked to the protein surface by a narrow channel. A large hydrophobic pocket was identified in the active site of HbHNL where the substrate is located. The channel is predominantly composed of apolar residues and Ser, Cys, and Thr, which are involved in oxyanion stabilization. The electrostatic potentials along the active site channels in HNL are neither positive nor negative, which is consistent with the view that all substrates and products of the HNL-catalyzed reaction are uncharged or close to neutral pH (Wagner et al., 1996, 811). It has been proposed that nucleophilic substitution of the protein by the cyanide ion leads to cyanohydrin formation (where acetone is the substrate) via a tetrahedral intermediate. The mutual orientation of the tetrahedral intermediate and the attacking nucleophile is such that the reaction should be chirality specific.

It has been pointed out that the enzyme-catalyzed cyanohydrin formation involves S_N^2-type nucleophilic attack by cyanide ion on the tetrahedral hemiacetal or hemiketal intermediate in contrast to the uncatalyzed cyanohydrin formation reaction in solution, where direct attack of the cyanide on carbonyl carbon occurs. A model of the tetrahedral intermediate of the natural substrate has been studied (Wagner et al., 1996, 811). The cyanohydrin forms by nucleophilic substitution by an incoming CN^-. The orientation of the tetrahedral intermediate formed is expected to be important for chiral preference. This orientation-dependent nucleophilic attack is a factor for chiral discrimination in the HbHNL.

Details of the active site structure of HbHNL concerning its influence on stereoselectivity about the chiral substrates such as mandelonitrile and 2,3-dimethyl-2-hydroxy-butyronitrile are studied based on the x-ray diffraction structure of enzyme–substrate complexes (Gartler et al., 2007, 87). Hydrogen bonds between the hydroxyl group and Ser80 and Thr11, and electrostatic interaction of the cyano group with Lys236 forms a network of interaction that is orientation specific and responsible for stereoselectivity. The arrangement of the active site residues around the mandelonitrile substrate is shown in Figure 5.2. The cyano group of the mandelonitrile forms favorable electrostatic interaction with the amino group of Lys236 and the imidazole of His235. The hydroxyl group of the mandelonitrile interacts with the hydroxyl group of Ser80 and hydroxyl group of Thr11. The phenyl ring of mandelonitrile interacts with the indole ring of Trp128 (Figure 5.3). While the foregoing favorable interactions are responsible for the preference toward the S-enantiomer, the mechanism of the exclusion of R-enantiomer cannot be understood by the space requirement within the active site. There is sufficient room for the placement of R-enantiomer (as generated by exchanging the –OH and hydrogen atom attached with the chiral center of S-mandelonitrile,

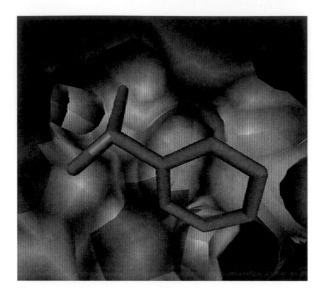

Figure 5.2 **See color insert.** The view of the active site and the channel linking to the protein surface of the hydroxynitrile lyase (1YB6.PDB) from the tropical rubber tree *Hevea brasiliensis* (Gartler, G. et al., 2007. *J. Biotechnol.*, 129: 87).

for example) within the active site. Rather, the exchange of the aforesaid groups will result in loss of favorable interactions that could be the potential cause of the chiral discrimination observed. A further possibility exists that one or more surrounding residues might be involved in stabilizing the transition state geometry of the S-enantiomer. The involvement of such residues is lost and raises the transition state barrier height in the case of R-enantiomer. Further electronic structure-based studies are required to address this issue. The hydrophobic pocket also specifies the attachment of the part of substrate by favorable interaction and augments the orientation-dependent interaction for the favored enantiomer and the active site (Gartler et al., 2007, 87). The active site as well as the channel are composed of predominantly apolar residues. Ser, Cys, and Thr are proposed to be involved in oxyanion stabilization. Site-directed mutagenesis showed that the replacement of Ser and His (by alanine) leads to the complete loss of enzyme activity. The replacement of Cys (by serine) also leads to loss of more than 95% catalytic activity. The site-directed mutagenesis studies of hydroxynitrile lyase from cassava (*Manihot esculenta Crantz*) (EC 4.1.2.37) by serine and histidine modifying reagents suggests involvement of active site seryl and histidyl residues (Wajant and Pfizenmaier, 1996, 25830).

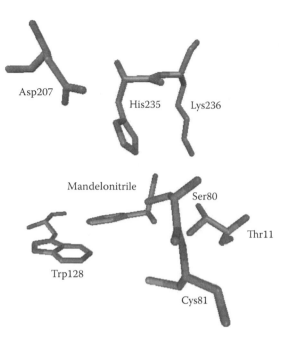

Figure 5.3 The active site residues (Thr11, Ser80, Cys81, Trp128, Asp207, His235, and Lys236) of the hydroxynitrile lyase (1YB6.PDB) from the tropical rubber tree *Hevea brasiliensis* around the mandelonitrile substrate (Gartler, G. et al., 2007. *J. Biotechnol.*, 129: 87).

5.2   Acceptance of both epimers of uronic acid by chondroitin lyase ABC

The family of bacterial chondroitin lyases ABC (chonABC) (EC. 4.2.2.20 and EC 4.2.2.21) has the feature that two different topologies of enzyme act on two different enantiomeric substrates. The chondroitin sulfate (CS) and dermatan sulfate (DS) are glycosaminoglycans, are linear, heterogeneous, and highly negatively charged polysaccharides composed of β-linked disaccharide repeating units containing uronic acid linked to *N*-acetyl-D-galactosamine. The uronic acid moiety in CS is exclusively β-D-glucuronic acid, whereas DS contains a mixture of α-L-iduronic acid and GlcA epimers. Both CS and DS degrade at the nonreducing end of either glucuronic acid or its epimer iduronic acid by a β-elimination mechanism. The enzyme with two different folds catalyzes the reactions for the individual enantiomers. The structure of chondroitinase ABC from *Proteus vulgaris* showed remarkable similarity with chondroitinase AC, for example, the presence of a Tyr-His-Glu-Arg catalytic tetrad. In rare cases, a single enzyme can process both enantiomers efficiently, but no detailed molecular understanding for such catalysis is available.

The structure of a *Bacteroides thetaiotaomicron* chondroitinase lyases ABC was studied to identify additional structurally conserved residues that might be potentially involved in catalysis (Shaya et al., 2008, 270). The authors searched for a distantly related member of the chonABC family, aiming to determine its structure and, by comparison with *Proteus vulgaris* chondroitinase ABC I (PvulABCI), to identify structurally conserved residues close to the catalytic tetrad, which are possibly involved in the catalysis of DS substrates. Another such enzyme, *Bacteroides thetaiotaomicron* chondroitinase ABC (BactnABC), with similar specificity as PvulABCI was identified in *Bacteroides thetaiotaomicron*. It has been suggested that the substrate-binding cleft may be narrowed in the presence of the substrate. A conserved cluster was identified in nanometer proximity from the catalytic tetrad. A histidine in this cluster is noted to be essential for the catalysis of DS but not CS. The enzyme utilizes a single substrate-binding site while having two partially overlapping active sites catalyzing the respective reactions. The spatial separation of the two sets of residues suggests a substrate-induced conformational change that brings close together all catalytically essential residues.

References

Daußmann, T., Rosen, T.C., and Dünkelmann, P. (2006). Oxidoreductases and hydroxynitrilase lyases: Complementary enzymatic technologies for chiral alcohols. *Eng. Life Sci.*, 6: 125–129.

Gartler, G., Kratky, C., and Gruber, K. (2007). Structural determinants of the enantioselectivity of the hydroxynitrile lyase from *Hevea brasiliensis*. *J. Biotechnol.*, 129: 87–97.

Griengl, H., Schwab, H., and Fechter, M. (2000). The synthesis of chiral cyanohydrins by oxynitrilases. *Trends Biotechnol.*, 18: 252–256.

Rosenthaler, L. (1908). Enzyme-effected asymmetric syntheses. *Biochem. Z.*, 14: 238–253.

Shaya, D., Hahn, B.-S., Bjerkan, T.M., Kim, W.S., Park, N.Y., Sim, J.-S., Kim, Y.-S., and Cygler, M. (2008). Composite active site of chondroitin lyase ABC accepting both epimers of uronic acid. *Glycobiology*, 18: 270–277.

Wagner, U.G., Hasslacher, M., Griengl, H., Schwab, H., and Kratky, C. (1996). Mechanism of cyanogenesis: The crystal structure of hydroxynitrile lyase from *Hevea brasiliensis*. *Structure*, 4: 811–822.

Wajant, H. and Pfizenmaier, K. (1996). Identification of potential active-site residues in the hydroxynitrile lyase from *Manihot esculenta*. *J. Biol. Chem.*, 271: 25830–25834.

chapter six

Chiral discrimination in the active site of ligases

The ligases are enzymes that catalyze a synthesis process (the joining of two molecules) with concomitant hydrolysis of the diphosphate bond in a triphosphate or ATP. This class of enzymes is also termed synthases, carboxylase, or synthetases. The ligases can be very specific in the chirality of the substrate. For example, aminoacyl-tRNA synthetase (aaRS) is a ligase that catalyzes the esterification of a specific amino acid and its cognate tRNA to form an aminoacyl-tRNA, leading to the charging of tRNA. The charging step is an essential prerequisite for protein synthesis, which depends on the chirality of the amino acid involved. The chance of misincorporation of wrong chirality is insignificant. Subsequently, a ribosome can transfer the correct amino acid from the tRNA onto a growing peptide chain, according to the genetic code, once the tRNA is charged with high fidelity. Chiral discrimination is rather stringent as developed by evolution. Similar to the different aaRS, other ligases show remarkable specificity in regard to the chirality of the substrate

6.1 Chiral discrimination by germacrene D synthases

Germacrene D is a naturally occurring, volatile sesquiterpene hydrocarbon with about 300 isomers. They are abundant in plants, microbes, and marine organisms, and are precursors of many biologically important compounds. They have strong effects on plant–insect interaction. It is known that the two enantiomers of germacrene D have different bioactivities. Germacrene D is an important intermediate in the formation of many sesquiterpenes. The (–) configuration is a more common enantiomer in higher plants. Synthesis of both enantiomers in *Solidago* species is controlled by two enantioselective synthases. It has been suggested that the two plant compounds are expected to give different signals to the animal and are expected to activate two different neuron types. The presence of one or two types of neurons is suggestive of the mechanism of chiral discrimination by a moth using germacrene D receptor neurons. The study of enantioselectivity of the germacrene D receptor neurons in *Helicoverpa armigera* indicated that all neurons, although they responded to both enantiomers, (–) germacrene D has about 10 times a stronger effect than (+) germacrene D (Stranden et al., 2004, 143). A parallel dose response curve for the two enantiomers is obtained by direct stimulation. It has been suggested that the two enantiomers mediate the same kind of message to moth via neuron response but with different intensity. It is suggested that the two enantiomers of germacrene D are transported by the same odor-binding protein and interact with the same membrane receptor protein, but with a different affinity (Stranden et al., 2004, 143). However, a detailed understanding of the interaction of the germacrene D enantiomers with the active site is yet to be carried out. Yamasaki and coworkers observed that (–) germacrene D acts as a masking substance of attractants for the cerambycid beetle, *Monochamus alternatus* (Yamasaki et al., 1997, 423). The study indicated that the locomotion of the female *Monochamus alternatus* is stimulated by the application of (+) juniperol and (+) pimaral, and inhibited by the action of (–) germacrene D. However, the study is not aimed at understanding the microscopic origin of chiral discrimination in odorant-receptor interaction.

A study of the interaction of the germacrene D enantiomers with the active site might be helpful in understanding the molecular basis of chiral discrimination. The study of the interaction between enantiomeric odorant molecules and peptide motifs (considered mimics of segments of the helical receptor structures) is available (Nandi, 2005, 1929). The model showed that the enantiomers of carvone have distinctly different interaction profiles with all peptides considered, but the enantiomers of camphor do not exhibit any significant enantiodifference. This result parallels the experimental fact that the enantiomers of carvone have different smells, while those of camphor cannot be discriminated by olfaction (Nandi, 2005, 1929). The recognition of the chiral odorant molecule (carvone) by a chiral lipid (dipalmitoyl phosphatidylcholine) is also theoretically studied, which explains the experimentally observed preference of the interaction of the lipid with one particular enantiomer of carvone over the other (Nandi, 2003, 588).

Two sesquiterpene synthase cDNAs (Sc11 and Sc19, respectively) are isolated from goldenrod (Prosser et al., 2004, 136). The functional expression in *E. coli* indicated that Sc11 is the (+)-germacrene D synthase, while Sc19 acts as (–) germacrene D synthase. Thus, different synthases having similarity in primary structure synthesize the enantiomers of germacrene D. Sc11 contains a motif as [303]NDTYD. Mutagenesis of this motif to [303]DDTYD leads to an enzyme that retained (+)-germacrene D synthase activity. The D303N mutation in Sc19 yields an enzyme with reduced activity. Mutation of 303 positions to glutamate in both Sc11 and Sc19 enzymes resulted in loss of activity. Very subtle changes to the active site of this family of enzymes are sufficient to change the reaction pathway leading to the formation of different enantiomers. As the enzymes show a remarkable degree of sequence identity, the study indicates that only very subtle changes in the active site architecture of the corresponding enzyme are required to switch the chirality-specific features of this particular enzymatic reaction. Although the biochemical studies clearly indicate that the active site structure is responsible for the experimentally observed chiral discrimination, microscopic studies are required to understand its molecular basis.

6.2 Chiral discrimination by aminoacyl-tRNA synthetases

The aminoacylation reaction involves amino acids (all of which are chiral molecules, except glycine), tRNA (containing chiral sugar rings), and ATP as substrates. The reaction occurs within the active site of aaRS, which is also composed of chiral amino acids. The active site of aminoacyl-tRNA synthetases is a cavity of nanodimension where these chiral moieties are confined, and it reacts in an orientation-specific way. Consequently, it is expected that the aaRSs would exhibit stringent chiral discrimination, and this has been confirmed experimentally. The specificity of the aaRSs in incorporating a specific amino acid and its ability to discriminate between several competing substrates is well known (Ibba and Söll, 2000, 617). Synthetase ensures that only cognate substrates are selected from the large cellular pool of similar amino acids or tRNA. The ability to discriminate among amino acids is one of the major requirements for accurate translation of the genetic code because the structure of the cognate and noncognate amino acids may differ only in a subtle way in terms of charge distribution or spatial arrangements of groups attached to the chiral center. The frequency of misincorporation of noncognate amino acid is 1 in 10,000, which reflects a surprising accuracy of the mechanism (Berg et al., 2002, 671). It was proposed earlier that there may be a succession of discriminating processes (Yamane and Hopfield, 1977, 2246; Hopfield, 1974, 4135). It has also been suggested that a number of intricate contacts between aaRS and amino acid, which are favorable for cognate but unfavorable for noncognate amino acids, makes possible successful recognition and concomitant discrimination (Ibba and Söll, 2000, 617).

Most of the aminoacyl-tRNA contains an editing site in addition to an acylation site. These complementary pairs of sites act as a double filter to ensure that noncognate amino acids are rejected. It is proposed in the literature that the acylation site rejects amino acids larger than the correct one because there is insufficient space for them. On the other hand, the smaller amino acids are cleaved by the hydrolytic editing site. However, this size- or volume-based mechanism is not capable of discriminating the enantiomeric species of the cognate amino acid as the D- and

L-enantiomers occupy identical total volume. Each synthetase can specifically recognize the correct pair of substrates from the pool of amino acids and tRNA molecules. The discrimination of cognate and noncognate amino acids is achieved by a number of intermolecular interactions between aaRS and substrates (amino acid and tRNA). Several workers have explored how a particular aaRS recognizes its cognate amino acid by various methods such as crystallographic analysis, biochemical methods, and mutation experiments, as well as computational analysis.

Although a detailed understanding of several aspects of the recognition of the substrates and mechanistic aspects of the reactions carried out by aaRSs is available from crystallographic and biochemical studies, studies on chiral discrimination are limited. Experimental studies indicate that the specificity is present at a level sufficient to discriminate the enantiomeric species. Early studies indicated that when D-isomer is tested as substrates with particle-free supernatant of pancreas homogenate containing synthetase, no activity is observed (Davie et al., 1956, 21). Calender and Berg investigated the substrate specificity of tyrosyl tRNA synthetases of *E. coli* and *Bacillus subtilis*, and noted that D-tyrosine was activated and transferred to tRNA (Calendar and Berg, 1966, 1690). A striking feature of a subsequent study is that identical quantities of L- and D-tyrosyl tRNA can be formed from L- or D-tyrosine, excess tyrosyl tRNA synthetase, and limiting tRNA (Calendar and Berg, 1967, 39). Partially purified preparations of alanyl-, leucyl-, and valyl-tRNA synthetases from *E. coli* showed D-enantiomers were esterified to tRNA at $<10^{-5}$ the rate of L-enantiomers. A barely detectable amount of D-phenyl alanyl tRNA was noted to be formed using phenylalanyl tRNA synthetases, and this was only $\sim 6 \times 10^{-5}$ the rate found for L-phenylalanyl tRNA. Tyrosyl tRNA synthetases from *E. coli* and *Bacillus subtilis* are unique in that they are the first enzyme that synthesizes D-aminoacyl-tRNA derivatives.

Soutourina et al. concluded that in addition to D-tyrosine, D-enantiomers of other amino acids could be incorporated based on an investigation *E. coli* and *Saccharomyces cerevisiae* systems (Soutourina et al., 2000, 32535). It has also been pointed out that deacylase activity might have a broader implication in retaining enantiopurity. The broad specificity of *E. coli* D-tyr-tRNA deacylase was early observed in vitro by Calendar and Berg (Calendar and Berg, 1967, 39). The deacylase recognizes very different D-aminoacyl moieties such as the acidic aspartate or the bulky aromatic tryptophan. The authors discuss another important pathway of removal of the toxic D-amino acid in addition to the proofreading mechanism during the aminoacylation step. In the case of auxotrophic bacteria, the lack of many biosynthetic pathways might result in low levels of endogenous D-amino acids. The deacylase activity would be redundant in such bacteria. However, D-aminoacyl-tRNA deacylase activity would be necessary in other cells to reduce the harmful effect of D-amino acid transfer to tRNA. Deacylase should be present to hydrolyze any D-aminoacyl-tRNA molecule. The authors have raised the question of why cells have maintained genes that encode a deacylase capable of hydrolyzing D-AA-tRNA^{Tyr} into free tRNA-AA and D-AA (AA being an amino acid such as Tyr) rather than evolving through the selection of more specific aminoacyl-tRNA synthetases. It has been suggested that the primitive cell might have indifferently incorporated L- and D-amino acids into polypeptides. At this stage, acquisition of a general D-aminoacyl-tRNA deacylase activity could have helped the cell to shift protein synthesis in the L-amino acid world. According to this situation, further evolution of cells implies an improvement of the specificity of the synthetases toward L-amino acids, and the progressive loss of a no longer useful deacylase gene, which becomes unnecessary. Cyanobacteria and archaebacteria may correspond to such cells. However, this hypothesis is yet to be confirmed.

Bergmann et al. tested the specificity of the L- and D-isomers of valine, isoleucine, and leucine for aminoacyl adenylate formation with valyl, isoleucyl, leucyl, and methionyl tRNA synthetases (Bergmann et al., 1961, 1735). In all cases, formation of the D-aminoacyl adenylate formation is negligible or nil compared to the corresponding L-aminoacyl adenylate formation. Similar chiral specificity is noted for aspartyl-tRNA synthetase of *Lactobacillus aerabinosus* (Norton et al., 1963, 269). Yamane and Hopfield observed that both L- and D-tyrosine can be esterified by tyrosine-tRNA synthetase (Yamane et al., 1974, 4135). This conclusion corroborates the study by Calendar and Berg. The amounts of ATP hydrolyzed per tRNA acylated are practically identical for both enantiomers. Tyrosyl tRNA synthetase is known to be nonspecific with respect to the acylating position (2′ or 3′ position of the hydroxyl group of the adenosine moiety), and it is suggested that it may lack the 2′ hydrolytic activity of other synthetases. This may also explain the existence of D-Tyr tRNA deacylase and absence of enzymatic deacylation of D-Tyr-tRNA (Yamane et al., 1974, 4135).

In a series of detailed studies, Hecht and coworkers demonstrated ways of preparation of tRNA activated with a variety of amino acids and amino acid analogues (Hecht, 1992, 545). The discrimination noted in the peptide synthesis using such misacylated tRNA (including D-amino acid) is also described. Profy and Usher studied the aminoacylation of diinosine monophosphate (inosinylyl-(3′-5′)-inosine), which is a model system with both 2′ and 3′ hydroxyl groups (Profy and Usher, 1984, 147). When the acylating agent was the imidazolide of N-(*tert*-butoxycarbonyl)-DL-alanine, a 40% enantiomeric excess of the L isomer was incorporated at the internal 2′ site. The positions of equilibrium for the 2′–3′ migration reaction differed for the D and L enantiomers. In contrast, reaction of IpI with the imidazolide of unprotected DL-alanine led to an excess of the D isomer at the internal 2′ site, while reaction with the n-carboxy anhydride of DL-alanine proceeded without detectable stereoselection.

Tamura and Schimmel studied the plausibility of chiral selectivity during the second step of aminoacylation by carrying out nonprotein, nonribozyme, RNA-directed aminoacylation of an RNA minihelix that recapitulates the domain within tRNA harboring the amino acid attachment site (Tamura and Schimmel, 2004, 1253). The minihelix[Ala] is constructed based on the sequence of *E. coli* tRNA[Ala], considered as a progenitor of modern tRNA. While the natural system uses aminoacyl phosphate (mononucleotide) adenylates as intermediates for aminoacyl-tRNA synthesis, the authors designed an aminoacyl phosphate oligonucleotide to hybridize to the 3'-end of the minihelix through a bridging oligonucleotide, thereby bringing together the activated amino acid and the amino acid attachment site. Minihelix[Ala], a bridging oligonucleotide, and 5'-[^{14}C]-L-Ala-p-dT$_6$dA$_2$ as well as 5'-[^{14}C]-D-Ala-p-dT$_6$dA$_2$ are mixed together to achieve aminoacylation of minihelix[Ala]. The 3' terminal A was replaced by either 2'-dA or 3'-dA. While the 2'-dA substrate was charged, the 3'-dA derivative was not. Thus, aminoacylation was specific for the 3'-OH. Formation of the [^{14}C]-L-Ala-minihelix is preferred over that of [^{14}C]-D-Ala-minihelix by a ratio of about 4:1. Both stereoisomers of Leu-p-dT$_6$dA$_2$ and of Phe-p-dT$_6$dA$_2$ were also tested for chiral selectivity. A clear preference for L- over D-leucine (or phenylalanine) was observed. No significant difference was seen when the spontaneous hydrolysis of 5'-[^{14}C] L-Ala-*p*-oligonucleotide versus 5'-[^{14}C] D-Ala-*p*-oligonucleotide was compared, either in the presence or absence of the bridging oligonucleotide. In addition, no significant difference in relative yields of L- versus D-aminoacyl-minihelix products was noted at 10, 20, or 30 min over the course of the reaction. Thus, chiral preference appears to occur during aminoacyl transfer from the 5'-phosphate to the minihelix. It might be noted that the mutual spatial orientation and distance between reacting molecular segments in the model system studied by Tamura and Schimmel could be different from a corresponding biological system, which might influence the observed ratio of selectivity.

Tamura and Schimmel also synthesized the aminoacyl phosphate oligonucleotide, bridging oligonucleotide, and minihelix[Ala], which contain L-ribose (in contrast with natural D-ribose). Minihelix[Ala], bridging oligonucleotide, and either 5'-[^{14}C]-L-Ala-p-dT$_6$dA$_2$ or 5'-[^{14}C]-D-Ala-p-dT$_6$dA$_2$ (both containing L-deoxyribose) were combined and analyzed for aminoacylation of minihelix[Ala]. Formation of [^{14}C]-D-Ala minihelix was preferred over that of [^{14}C] L-Ala minihelix by a ratio of 1:3.6, about the reciprocal of that determined when RNA-containing D-ribose is used. This result again points out the complementary heterochiral relationship between L-amino acid and D-sugar, which seems to also play a role in the aminoacylation reaction. The chiral selectivity of the reaction with three different amino acids (alanine, leucine, and phenylalanine) was fourfold; although small, it can lead to an overwhelming preference for an L-amino acid in a biological system.

In a subsequent study, Tamura and Schimmel attempted to address whether the free amino group attached to the asymmetric α-carbon of the amino acid plays a role in the chiral selectivity of aminoacylation (Tamura and Schimmel, 2006, 13750). The aminoacylation reaction was performed with minihelix^{Ala}, bridging oligonucleotide and 5'-n-acetyl-[^{14}C]Ala-p-dT$_6$dA$_2$. It has been pointed out that free amino group (whether or not protonated) of amino acid as well as a free ribose hydroxyl are possible candidates to interact with a phosphate oxygen in the process of formation of the relevant transition state of the reaction. To address the question, the amino group of Ala was acetylated, and n-acetyl-L- or D-Ala-aminoacyl oligonucleotide was used for the aminoacylation reaction. Acetylation of Ala was performed with more than 70% of both L-[^{14}C]Ala and D-[^{14}C]Ala being acetylated. The ratio of the product formed (calculated from the band intensities of products resolved by gel electrophoresis) after 30 min at 0°C was 3.7:1 (Ac-L-Ala: Ac-D-Ala). This selectivity was similar to that for nonacetylated Ala, which was 1:4. These results suggest that, at least during the rate-determining step, the amino group of Ala does not make any hydrogen bonds with the phosphate oxygen of the terminal adenosine of the minihelix or with another H-bond acceptor located elsewhere in the complex.

Consideration of steric models of the base pairing at the position closest to the amino acid attachment site shows that a potential clash of the CH$_3$ of dT with the CH$_3$ of L-Ala is absent in the Watson–Crick dT:A pair (Tamura and Schimmel, 2006, 13750). While the CH$_3$ of D-Ala crowds the 3'-OH of A, the CH$_3$ of L-Ala is distal to the same 3'-OH. Based on such model building, a dT:G pair was introduced to create a potential clash of the CH$_3$ of L-Ala with that of dT. This substitution sharply reduced the yield of L-Ala-minihelix^{Ala} without altering the production of D-Ala-minihelix^{Ala}. A suppression of the chiral-selective aminoacylation was noted and even somewhat reversed in favor of the D-Ala-minihelix^{Ala} product. Similarly, removal of the ring CH$_3$ through substituting at dU: G for a dT:G pair, gave a structure in which chiral preference for L-Ala was retained, which is consistent with the possibility of a clash between the ring and L-amino acid methyl groups. They also investigated the subtle effect of steric clash due to sugar ring puckering in the process of selectivity. In the Watson–Crick dT:A pair, the potential clash of the CH$_3$ of dT with the CH$_3$ of L-Ala could be avoided through a 3'-endo pucker of dT. In contrast, conversion to the 2'-pucker could bring the CH$_3$ of dT close to the CH$_3$ of L-Ala. The NMR structure of a dT:dG base pair established a 2'-endo preference for the deoxyribose of dT. A 3'-endo pucker preference can be created by making a ribose 2'-O-CH$_3$ substitution. Accordingly, a dT (2'-O-Me):G pair was introduced, and chiral-selective aminoacylation in favor of the L-Ala product was observed. Also, to remove the capacity for making a water bridge between the 3'-O and a base that would stabilize the 2'-endo conformation, a dT:I (inosine) pair was also tested. This construct also showed the chiral preference for L-Ala. These results support the idea that the chiral preference for L-Ala in these constructs depends on avoiding a sugar-pucker sensitive steric clash between a pendant group of a base with the CH_3 of L-Ala.

Tamura and Schimel pointed out that the chiral preference can arise during the process of aminoacyl transfer from the 5'-phosphate of the oligonucleotide to the minihelix or from a difference in template (bridging oligonucleotide) hybridization efficiency between the L-amino acid- and D-amino acid-poligonucleotides, rather than during any other reaction within the ternary complex itself. The consequences of varying the concentration of the bridging oligonucleotide are explored to determine whether the rate of the aminoacylation was sensitive to its concentration and at the same time affected the ratio of the L- to D- products. The amount of aminoacylation measured at four different time points was sensitive to the concentration of bridging oligonucleotide. However, at all time points and at all concentrations of the bridging oligonucleotide, the nearly four-fold preference of L-Ala-minihelix[Ala] to D-Ala-minihelix[Ala] was observed. This result further supports the idea that chiral selectivity comes during aminoacyl transfer in the tripartite complex. The detailed work of Tamura and Schimmel indicates that an energetic difference of <1 kCal.mol^{-1} in the rate-determining step of the transition state is sufficient to give the observed fourfold preference for chiral specificity in the aminoacylation process. It may be noted that the ease of formation of a transition state in the model tripartite complex studied by the authors is greater than with the corresponding reaction within the active site of aminoacyl transferase. The reacting moieties in the latter case are surrounded by the active site residues, which influence the progress of reaction in the cases of L- and D- reactant.

The remarkable specificity of the reaction must be dependent on the architecture of the active site of aaRSs. The active site residues not only form a perfectly complementary binding pocket for the substrates but also carry the reactants to the product state via the transition state. In the following text, we discuss features of the construction of the active site in different aaRSs, which seems useful for tracking the generic roles of various conserved residues in the reaction mechanism. The active site cavity of aaRS can be subdivided into three regions: the ATP binding pocket, the amino acid binding pocket, and the intervening region, which is the region closest to the reaction center (the region close to the ester linkage) (Rath et al., 1998, 439). The schematic representations of the three regions of the active site of HisRS of prokaryotic *E. coli* histidyl-tRNA synthetase generated from the crystal structures 1KMN.PDB (Arnez et al., 1997, 7144)

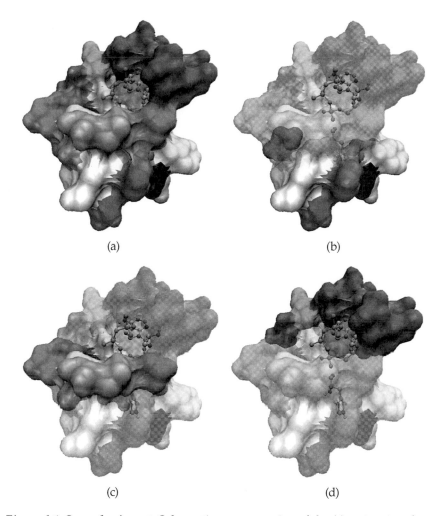

      (a)   (b)

      (c)   (d)

Figure 6.1 **See color insert.** Schematic representation of the (a) active site of pro-karyotic *Escherichia coli* histidyl-tRNA synthetase (1KMN.PDB) (Arnez et al., 1997. *Proc. Natl. Acad. Sci. (USA)*, 94: 7144), which includes the amino acid binding site, ATP binding site, and intervening region; (b) amino acid binding pocket; (c) inter-vening region; and (d) ATP binding pocket. The reactants (histidinol, an inhibitor and ATP) are shown by ball and stick representation, and the surrounding active site cavity is shown by surface representation. The images are prepared using VMD. (Humphrey et al., 1996: 33.)

are shown in Figure 6.1. Since the active site structures of different aaRSs are designed by evolution to recognize its cognate amino acid as well as ATP, it is instructive to look into the nearest-neighbor residues in close proximity to each of the reactants. The residues in the three pockets have close interaction with the substrates during the course of the reaction.

The molecules involved in the aminoacylation reaction are ATP, amino acid, and tRNA. The molecular fragments of ATP and amino acids are (1) the adenosine group; (2) the ribose sugar group; (3) the triphosphate group bearing a negative charge (belonging to ATP); (4) positively charged amino groups of the substrate amino acids in zwitterionic form; (5) the negatively charged carboxylic acid group in the zwitterionic form of the amino acid; and (6) the side chain (except glycine), which differs in polarity, charge, or hydrophobicity for 20 amino acids and may contain a functional group (positively or negatively charged or polar). In corroboration with the chemical structure of the foregoing groups of the substrates, the active site residues of the corresponding aaRSs have complementary chemical structure so that favorable interactions can bind the substrates effectively and with fidelity. This complementary nature of the interaction between substrate and active sites is also responsible for carrying out the reactants to form the product via a transition state through the reaction pathway. The pattern of the interaction depends on the chiral structure of each of the active site residues and substrates. Hence, it is not surprising that chiral discrimination is significantly observable in this reaction. Below, we refer to the chemical structure of the substrates and various active site residues of the corresponding aaRSs that have complementary favorable interactions. The various species for which the information on the active site of different aaRSs is available are *Bacillus stearothermophilus* (BS), *E. coli*, *Thermus thermophilus* (TT), and *Saccharomyces cerevisiae* (yeast).

The π electron cloud present in the adenosine group of ATP can have favorable π-stacking interaction with the electron cloud of active site amino acids such as Phe or His. The presence of such residues is noted in different aaRSs. Active site His is present in ArgRS (Newberry et al., 2002, 2778) and in GluRS (Sekine et al., 2003, 676), which interacts with the adenosine group. Interaction of the π electron cloud of active site Phe group with the adenosine group is noted in AsnRS (Colominas et al., 1998, 2947), SerRS (Biou et al., 1994, 1404, Belrhali et al., 1995, 341), and HisRS (Arnez et al., 1997, 7144) as well as in GlyRS (Arnez et al., 1999, 1449). Further hydrogen bonding between the nitrogen atoms of the adenosine base and active site residues such as His, Pro, Asp, Asn, Met, and Gln are observed. Such hydrogen-bonding interactions are present in IleRS with active site His (Nakama et al., 2001, 47387; Osamu et al., 280, 1998) and with Met in ValRS (Fukai et al., 2000, 793).

The reduction of the unfavorable electrostatic potential between the ATP and amino acid (due to the presence of negative charge density over the triphosphate group of ATP, as well as the carboxylic acid group of amino acid) is a prerequisite to carry out the reaction effectively. The positively charged amino acids (such as His, Lys, and Arg, and polar residues such as Thr or one or more divalent Mg^{2+} cation) carry out the task of reduction of electrostatic potential. The presence of one or more such residue or ion is observed in the ATP binding site of different aaRSs. Interactions with His residues in the active site and triphosphate group of ATP are observed in ArgRS (Benedicte et al., 2000, 5599; Cavarelli et al., 1998, 5438) and CysRS (Newberry et al., 2002, 2778). Similarly, interactions with the triphosphate group of ATP and different positively charged amino acid residues and cation are noted in GlnRS with His and Lys (Rath et al., 1998, 439; Perona et al., 1993, 8758), in GluRS with Arg (Sekine et al., 2003, 676), IleRS with His, Lys (Nakama et al., 2001, 47387; Nureki et al., 1998, 280), TyrRS with Lys (Yaremchuk et al., 2002, 3829; Xin et al., 2000, 299; Osamu and Fersht, 1988, 1581), and in AsnRS with Arg and His residues as well as three Mg^{2+} ions (Colominas et al., 1998; 2947). The triphosphate group of ATP in ProRS interacts with active site Arg, His, and Thr (Yaremchuk et al., 2001, 989). Interactions between Arg residues and Mg^{2+} ions, and the triphosphate group of ATP are noted in class II aaRSs such as GlyRS, HisRS, LysRS, and SerRS (Arnez et al., 1999, 1449; Arnez et al., 1997, 7144; Desogus et al., 2000, 8418; Onesti et al., 2000, 12853; Biou et al., 1994, 1404; Belrhali et al., 1995, 341). The triphosphate groups are also involved in hydrogen-bonding interaction with polar active site groups in some aaRSs. In few cases, the triphosphate group is involved in hydrogen bonding with amide hydrogen in a main chain of a peptide linkage located nearby in the active site.

The positively charged α-amino group of the substrate amino acid can have favorable interaction with negatively charged active site residues such as Asp or Glu. The negatively charged active site Asp residue interacts with positively charged α-amino group of the substrate amino acid in GlnRS (Rath et al., 1998, 439; Perona et al., 1993, 8758) and in LeuRS (Cusack et al., 2000, 2351). Similarly, interaction between the negatively charged active site Glu and the α-amino group of the substrate amino acid is noted in GluRS (Sekine et al., 2003, 676), in LysRS (Desogus et al., 2000, 8418; Onesti et al., 2000, 12853), SerRS (Biou et al., 1994, 1404; Belrhali et al., 1995, 341) and in HisRS (Arnez et al., 1997, 7144). Active site residues with polar side chains such as Asn, Ser, Gln, Tyr, and Thr are also observed in the vicinity of the α-amino group of substrate amino acid, which can form stabilizing hydrogen bond interaction. The presence of such a hydrogen bond is noted between α-amino groups of the substrate amino acid and Asn as well as Ser in ArgRS (Benedicte et al., 2000, 5599; Cavarelli et al., 1998, 5438) and with Thr in CysRS (Newberry et al., 2002, 2778).

The α-carboxylic acid group of the substrate amino acid can have favorable interaction with the positively charged active site amino acid residues such as Arg or Lys, His, or active site amino acids containing polar side chains such as Asn or Gln. Various such active site residues bearing a positive charge or of polar nature, present in the proximity of the α-carboxylic acid group of the substrate amino acid of different aaRS active sites, are noted. The examples are residues such as Asn, Gln, and

His in ArgRS (Benedicte et al., 2000, 5599; Cavarelli et al., 1998, 5438), Arg and Lys in LysRS (Desogus et al., 2000, 8418; Onesti et al., 2000, 12853), Arg in HisRS (Arnez et al., 1999, 1449; Arnez et al., 1997, 7144), Gln in IleRS (Nakama et al., 2001, 47387; Osamu et al., 280, 1998), and His in LeuRS (Cusack et al., 2000, 2351). Other than the polar groups (as mentioned before), the amide group of a peptide linkage can form a hydrogen bond with the carboxylic acid group.

Unlike the invariant carboxylic and amino groups, the side chains are variable in the structure of amino acids (glycine has no side chain). For example, the side chains could be positively charged, negatively charged, polar, nonpolar, or may have a sulfur-containing group. When the side chain of the substrate amino acid is positively charged similar to those in Arg or Lys, the vicinity of the active site close to the side chain is composed of negatively charged amino acids such as Glu, Asp, or polar amino acid such as Tyr. Presence of active site Glu, Asp, and Tyr is noted in the proximity of positively charged side chain of the substrate Arg (Benedicte et al., 2000, 5599; Cavarelli et al., 1998, 5438). Active site residues Tyr and Glu are present in LysRS (Desogus et al., 2000, 8418; Onesti et al., 2000, 12853). Similar interactions with Tyr and Glu are noted in HisRS, where the side chain of the substrate amino acid can be in the protonated state (ionic) or nonprotonated (polar) state (Arnez et al., 1997, 7144).

The binding pocket for the substrate amino acid bearing negatively charged side chain (such as Glu, Asp) may contain a positively charged active site amino acid such as Arg or Lys, or an amino acid that can form a hydrogen bond (such as Gln, Tyr, Ser, or Asn) with the carboxylic acid group of substrate amino acid side chain. In GluRS, Arg, Tyr, Asn, and Arg interact with the side chain (Sekine et al., 2003, 676). In AspRS, interactions with the active site Arg, Lys, Gln, Asp, and Ser are noted in the vicinity of the negatively charged side chain of substrate amino acid (Eiler et al., 1999, 6532; Cavarelli et al., 1994, 327).

When the side chain of the substrate amino acid is polar (for example, side chains of Gln, Asn, Ser, Cys), hydrogen bonding with polar active site residues such as Tyr, Thr, or water molecule or interactions with charged active site residues such as Glu and Arg are observed. Such interactions with Tyr and water molecule are noted in GlnRS (Rath et al., 1998, 439; Perona et al., 1993, 8758), with Arg and Glu are noted in AsnRS (Colominas et al., 1998, 2947), and with Thr and Glu are noted in SerRS (Biou et al., 1994, 1404; Belrhali et al., 1995, 341).

In the case of substrate amino acid with a nonpolar side chain, the corresponding active site pocket is composed of nonpolar amino acids such as Pro, Trp, Phe, and Ile, which might have favorable van der Waals and hydrophobic interactions with the side chain of the substrate. Further, the presence of the hydrophobic part of the side chain of amino acids such as Met, His, and Tyr are observed near the nonpolar side chain of the substrate. Examples of the nonpolar active site residue and the hydrophobic part of the side chain of charged or polar amino acids near the nonpolar side chain of the substrate amino acid are as follows. Pro and Trp are present in the active site of IleRS (Nakama et al., 2001, 47387; Osamu et al., 280, 1998); Pro, Ile, and Trp are present in ValRS (Fukai et al., 2000, 793); Trp and Gly are present in ProRS (Yaremchuk et al., 2001, 989); and Met, Phe, Tyr, and His are present in LeuRS (Cusack et al., 2000, 2351) near the nonpolar side chain of the substrate amino acid.

Of course, the generic interactions mentioned before are not limited to the foregoing types. Long-range influences of the distant residues (located in the second shell around the active site or beyond) present in enzyme must be there due to the long-range nature of electrostatic interactions. It is, however, expected that chiral discrimination will largely be influenced by nearest-neighbor interactions. Moreover, in the cases of a few amino acids such as Cys or Gly, customized interactions by active site residues or ions to form favorable interactions with these amino acids are noted. The side chain-binding pocket Cys contains a Zn^{2+} ion, which is involved in a coordination interaction. The Zn^{2+} ligates the Cys substrate via the side chain thiolate, while the remaining four coordination sites of Zn^{2+} ion are satisfied by the active site amino acids such as multiple Cys, His, and Glu (Newberry et al., 2002, 2778). Gly is the smallest amino acid and is achiral. Due to the absence of a side chain, the specificity of GlyRS depends exclusively on the carboxylic acid and amino acid group. The negatively charged carboxylic acid group of Glu88 and Glu239 interacts directly with the α-NH^{3+} group. Glu359 fills up the space in the binding pocket that could have been occupied by the side chain of substrate amino acids other than Gly and helps in retaining the fidelity exhibited by GlyRS. The pro-L α-hydrogen atom interacts with the carboxyl oxygen atom of Glu359. The side chain oxygen atom of serine is involved in the hydrogen bonding with the α-ammonium group. This interaction specifies the orientation of substrate glycine moiety and prevents binding of amino acid with similar chemical structure such as alanine by creating a steric block for the methyl group of alanine (Arnez et al., 1999, 1449).

We summarized earlier the presence of a few typical favorable interactions of ATP and substrate amino acids with conserved active site residues of different aaRSs based on the available crystallographic data. The foregoing data reveals that the substrate is located in a pattern of interaction (principally electrostatic such as ionic, dipolar, hydrogen bonding, or van der Waals and hydrophobic interaction). Since the active site residues are chiral with the exception of Gly, the resulting interaction patterns are different for each aaRS, which is responsible for the chiral discrimination. While the network of interaction is favorable for the chiral structure of the cognate substrate amino acid, it is unfavorable for a noncognate amino acid. In the latter case, one or more such interactions are absent or become unfavorable or even become repulsive for the noncognate substrate. Since the electrostatic nature of the groups of the substrate amino acid and ATP are limited to polar, nonpolar, positively or negatively charged or groups with π electron cloud, the interaction patterns with these moieties and active site residues also bear commonalities in terms of the interactions mentioned earlier. Presumably, the residues forming such common interaction patterns for an amino acid are conserved through evolution in different species.

Explicitly, the interaction between the active site residues of a given aaRS and its cognate amino acid might be categorized as substrate binding (ionic interaction, hydrogen bonding, and hydrophobic interaction), charge neutralization, and catalytic action. Some residues close to the reaction center can also have multiple roles. These functional roles are expected to be played by one or more active site residues of each of the different aaRSs, and are distributed through the three pockets on the basis of their function.

Crystallographic, biochemical, and kinetic studies clearly indicate that the chiral discrimination in the first and second steps of aminoacylation is stringent. However, the molecular origin of the chiral discrimination was not clear. The size-based mechanism suggested in the literature (Ibba and Söll, 2000, 617) is not capable of explaining the observed discrimination in the enantiomeric species of the cognate amino acid, as the D- and L- enantiomers have identical total size. Neither the acylation site nor the editing site would be able to discriminate the enantiomers if the size factor is essentially controlling the proofreading process. The chiral discrimination by aaRSs can be better understood in terms of the difference in electrostatic interactions between aaRS and amino acid. While this view is consistent with the studies of Warshel and coworkers (Warshel et al., 2006, 3210), it is further supported by specific studies made in the case of AspRS studied by Archontis et al. (Archontis et al., 1998, 823). The authors studied how aspartyl-tRNA synthetase distinguishes between negatively charged aspartic acid and neutral asparagines using molecular dynamics simulation and free energy simulation. This computational study elucidates that the analysis of electrostatic interaction would be useful to understand the discrimination between a cognate and noncognate amino acid. To correctly account for the electrostatic interactions in the system (including bulk solvent), the region of the mutation site is treated microscopically, whereas distant protein and solvent are treated by continuum electrostatics. The substrate Asp is predicted to bind much more strongly than Asn, with a significant binding free energy difference. This implies that erroneous binding of Asn by AspRS is very unlikely. Almost all of the protein contributions to the Asp versus Asn binding free energy difference arise from an arginine and a lysine residue that is hydrogen-bonded to the substrate carboxylate group, and an Asp and a Glu that are hydrogen-bonded. These four amino acid residues are completely conserved in AspRSs. The protein effectively solvates the Asp side chain more favorably compared to the aqueous solvation of the same side chain. Some synthetases discriminate between amino acids by forming a network of interactions between the amino acid and its specificity pocket (Fersht, 1987, 8031; Fersht and Dingwall, 1979, 1245; Fersht, 1986, 69).

Recent computational studies provide an understanding of the chiral discrimination in different aaRSs. Thompson et al. investigated the preference for acylation of L-Asp over D-Asp by aspartyl-tRNA synthetase (AspRS) using detailed classical molecular dynamics simulations and considering the electrostatic interactions present in the system. The study points out that the influence of the network of electrostatic interaction present in aaRS protects AspRS against most binding errors. Inverted D-Asp binding gives a free energy higher than L-Asp. Discrimination against D-Asp can be explained by the unfavorable binding with the D-Asp. It has been pointed out that the inverted D-Asp binding is slightly better than D-Asp binding in the regular orientation. It is predicted that AspRS is strongly protected against inverted L-Asp binding. Based on component analysis, it has been indicated that Coulomb interaction with the protein is the major favorable factor to the L-Asp binding. It has also been concluded that the co-binding of ATP and its three divalent cations does not significantly affect binding specificity. The model system includes the synthetase, amino acid, ATP, co-bound ions, and the solvent molecules (Thompson et al., 2007, 30856). It has been pointed out that the proofreading of Asp is difficult for AspRS, as it must protect not only against D-Asp, but against an inverted orientation where the two substrate carboxylates are swapped. Inverted D-Asp binding gives a higher free energy than L-Asp. discrimination against D-Asp can be explained by unfavorable binding with the inverted D-Asp where binding is slightly better than D-Asp binding in the regular orientation. This raises the possibility that D-Asp may acylate tRNA with its side chain rather than its backbone carboxylate (at least partly). For both TyrRS and AspRS, a moderate binding free energy difference between the L- and D-amino acids is noted, which is in agreement with their known ability to misacylate their tRNAs. It is predicted that AspRS is strongly protected against inverted L-Asp binding. Based on component analysis, it is indicated that Coulomb interactions with the protein provide +13.1 k.Cal. mol^{-1} in favor of the L-Asp geometry, offset by −6.2 k.Cal. mol^{-1} from water. An additional +3.2 k.Cal. mol^{-1} energy is contributed by van der Waals interactions, giving a total free energy of +10.4 k.Cal. mol^{-1} in favor of the regular L-Asp binding geometry. Thus, less favorable van der Waals interactions for inverted L-Asp contribute 31% of the overall effect, whereas Coulomb interactions contribute 69 k.Cal. mol^{-1}. This is consistent with the presence of an intricate network of electrostatic interaction present in and around the active site. As mentioned before, co-binding of ATP and its three divalent cations does not affect binding specificity and contributes little to the total free energy. The weaker coordination of the displaced ligand ammonium to the flipping loop Glu171 prevents L-Asp from binding in an inverted geometry. The study points out that the networks of electrostatic interaction present are the major influence that protects AspRS against most binding errors.

Electronic structure-based studies are aimed at understanding the chiral discrimination exhibited by the synthetases. Chiral discrimination has been studied in the first step of the aminoacylation reaction by HisRS using combined quantum mechanical:molecular mechanical calculation. A two-layered ONIOM model is used (*ab initio*: semiempirical level of theory) to study the discrimination between L-His and D-His, using a model of the corresponding active site (Dutta Banik and Nandi, 2009b, 468). To construct the model of the active site, the crystal structure of the oligomeric complex of histidyl-tRNA synthetase (HisRS) from *E. coli* complexed with ATP and histidinol and histidyl-adenylate (Arnez et al., 1997, 7144) was used. In this initial study, it was pointed out that the His is located in a network of interactions within the nanospace enclosed by the active site residues, which is perturbed when the D-amino acid is incorporated. The pathway to the transition state corresponding to the activation by D-His is also unfavorable due to the lack of stabilizing interactions in the transition state structure corresponding to the activation by L-His. The approach of L-His is easier within the active site than D-His during the course of the reaction (Dutta Banik and Nandi, 2009b, 468). Comparison of the reactant and the product geometry shows that the reacting moieties (amino acid and ATP) should undergo a change in separation and orientational rearrangement, which contributes to the discrimination. The corresponding distance-dependent interaction profile is shown in Figure 6.2. The variation in the energy is more restricted and unfavorable for the D-His compared to L-His, which is a signature of chiral discrimination. The restriction of the degrees of freedom enhances the chiral discrimination. The reorientation required for going to the product state for L-amino acid is without any significant energy barrier, and its approach is favored relative to the same for the corresponding D-amino acid. The result indicates that the incorporation of the D-form as an adenylate in the first step is strongly unfavored. This preliminary study pointed out different factors as responsible for the discrimination and the high level of stereospecificity.

At least four factors are responsible for the observed chiral discrimination. First, the distance and orientational changes involved in the approach of D-amino acid toward the ATP is unfavorable within the active site. Second, the charge distribution is important for discrimination and the removal of the charges in the model drastically reduces the

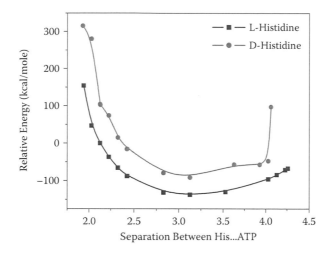

Figure 6.2 The variation of the relative intermolecular energy (relative to the energy in the model based on crystal structure expressed in kCal.mol^{-1}) as a function distance (in Å) of the L- and D-amino acid moiety measured between the oxygen of the carboxylic group of the His moiety and the α-phosphorus atom of ATP studied in a model of the active site. The Gln-83, Arg-113, Gln-127, Glu-131, Arg-259, and Tyr-261 are present as surrounding residues in the model. (Dutta Banik, S. and Nandi, N. 2009a, b. *Coll. Surf. B: Biointerfaces* 74: 468; *J. Surf. Sci. Technol.*, 25: 1. Reprinted with permission from Elsevier.) All calculations are performed by Gaussian 03W suite of programs (Frisch, M J et al., 2004. *Gaussian 03, Revision C.02*, Wallingford, CT: Gaussian, Inc.).

discrimination. The third factor contributing to the discrimination is the restricted nature of the mutual orientation within the cavity of the active site where the His and ATP are located. Due to the restricted nature of the change in orientation during the adenylate formation, the related interaction profiles are different for L-His and D-His. Fourth, the stereochemistry of D-His is such that it destabilizes the transition state due to unfavorable electrostatic interaction with surrounding residues.

In a subsequent study, the reaction mechanism of the first step of aminoacylation and the origin of discrimination was studied in further detail. The mechanism of activation of His has been proposed based on detailed analysis of the crystal structure of *Thermus thermophilus* histidyl-tRNA synthetase (Åberg et al., 1997, 3084) and *E. coli* histidyl-tRNA synthetase (Arnez et al., 1995, 4143). An in-line displacement mechanism, followed by the nucleophilic attack of the carboxyl group of the amino acid on the α-phosphate of the ATP, leads to the formation of histidyl-adenylate, which is a stable enzyme bound intermediate is suggested based on structural analysis. The electronic structure-based study of Dutta Banik and Nandi confirms many of the conclusions based on structural and

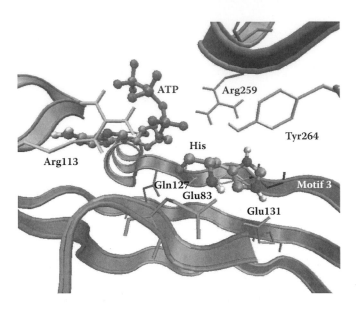

Figure 6.3 **See color insert.** Active site of the HisRS-Histidinol-ATP complex (1KMN.PDB) from *E. coli.* The reactants (histidinol and ATP) are shown in CPK style. Side chains of the active site in close proximity to the reactants as Glu 83, Arg 113, Gln 127, Glu 131, Arg 259, and Tyr 264 are shown by bonds. The remaining part of the motif 2, His A loop, and motif 3 are represented by ribbons. (Dutta Banik, S. and Nandi, N. 2010. *J. Phys. Chem. B.,* 114: 2301.) The image is prepared using VMD. (Reprinted with permission from the American Chemical Society. © American Chemical Society.)

biochemical methods (Dutta Banik and Nandi, 2010, 2301). The fidelity mechanism for the preference of activation by L-His over D-His in HisRS is studied using combined *ab initio*/semiempirical method. The model of the active site is shown in Figure 6.3. The change in orientation between His and ATP in going from the reactant to the product state is shown in Figure 6.4. The variation in the energy during the mutual approach of the His and ATP to form adenylate as a function of coordinates (which are changing during the progress of the reaction) using optimized geometry of L-His (denoted by Model$^{Opt}_{Reactant\ (L)}$) within the model of the active site is studied. The result shows that the surrounding nanospace of synthetase confines the L-His and ATP in the global minima of the plot and proximally places the reactants in geometry suitable for the in-line nucleophilic attack (Figure 6.5a). The incorporation of electron correlation using a two-level ONIOM (MP2/6-31G**:PM3//HF/6-31G**:PM3) does not change the nature of the potential energy surface as shown in Figure 6.5b.

   The energy surface becomes highly unfavorable when the D-His is placed within the active site (shown in Figure 6.6a and indicated as

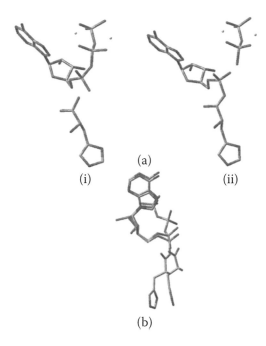

(a)

(i)   (ii)

(b)

Figure 6.4 (a) Comparison of the (i) reactant and the (ii) product as obtained from reactant and product crystal structure (1KMN.PDB and 1KMM.PDB, respectively) of HisRS. A change in the orientation of the amino acid moiety relative to the ATP occurs as shown in a(i) and a(ii), respectively. (b) The overlay of reactant and product structures indicating the displacement of the imidazole group of the His moiety toward the plane containing the sugar ring of the ATP coupled with a change of the orientation of the carboxylic acid group of the His in going from reactant state to the product state (Dutta Banik, S. and Nandi, N. 2010. *J. Phys. Chem. B.*, 114: 2301). (Reprinted with permission from the American Chemical Society. © American Chemical Society.)

Model$^{Unopt}_{Reactant\ (D)}$). The unfavorable nature of the potential energy surface of D-His becomes substantially favorable when the structure is reoptimized and the surrounding residues are rearranged to lower the interaction energy between the residues and D-His (indicated as Model$^{Opt}_{Reactant\ (D)}$ shown in Figure 6.6a). However, the rearranged geometry is still higher in energy compared to the structure containing L-His (shown in Figure 6.6b). This indicates that the arrangement of the nanospace of the active site of synthetase to allow the incorporation of D-His is ruled out by the cost of energy to be paid for such misincorporation. This indicates that the active site of the synthetase needs large-scale structural reorganization in order to reduce the unfavorable nature of the energy landscape of the model of the active site including D-His. Large-scale structural rearrangement of the peptide linkage of different motifs and

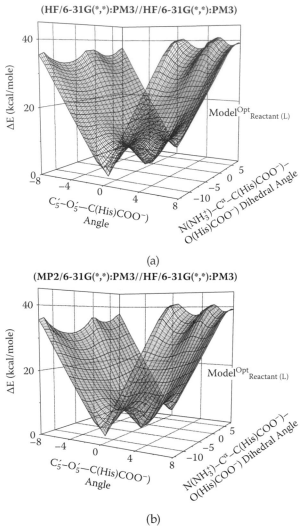

(a)

(b)

Figure 6.5 The variation of the interaction energy as a function of the orientation of the His moiety (expressed in degrees) and the orientation of the carboxylic acid group of L-His relative to the ATP (expressed in degrees) in the presence of active site surrounding residues as studied using optimized geometry (denoted by Model^{Opt}$_{Reactant (L)}$). The His is orientated (expressed in degrees) in a plane containing the C5′-O5′—C(COO⁻His) atoms (nearly perpendicular to the plane containing the imidazole group of the His). Simultaneously, the orientation (expressed in degrees) of the carboxylic acid group of His is varied with respect to the α–NH⁺₃—group of His. The variation in the energy is computed using (a) two-level ONIOM (HF/6-31G**:PM3// HF/6-31G**:PM3) and (b) the variation of the interaction energy as shown in (a) using a two-level ONIOM (MP2/6-31G**:PM3//HF/6-31G**:PM3) method. (Dutta Banik, S. and Nandi, N. 2010. *J. Phys. Chem. B.,* 114: 2301. Reprinted with permission from Elsevier.)

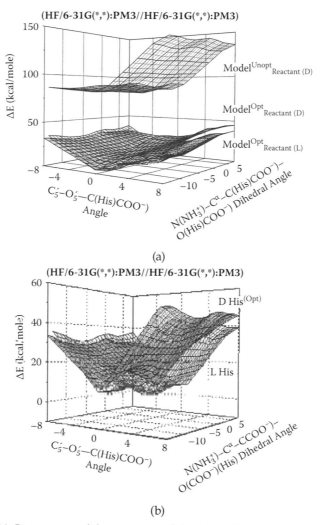

Figure 6.6 (a) Comparison of the variation of the interaction energy as a function of the orientation of the His moiety (expressed in degrees) and the orientation of the carboxylic acid group of His relative to the ATP (expressed in degrees) for $\text{Model}^{\text{Opt}}_{\text{Reactant}}$ (L), optimized model containing D-His is placed within the active site ($\text{Model}^{\text{Opt}}_{\text{Reactant}}$ (D)) and model containing D-His is placed within the active site without optimization ($\text{Model}^{\text{Unopt}}_{\text{Reactant (D)}}$). The variation in the energy is computed using two-level ONIOM (HF/6-31G**:PM3//HF/6-31G**:PM3) (b) Comparison of the variation of the interaction energy as shown in (a) with the interaction energy difference between $\text{Model}^{\text{Opt}}_{\text{Reactant}}$ (L), $\text{Model}^{\text{Opt}}_{\text{Reactant (D)}}$ in an enlarged scale (Dutta Banik, S. and Nandi, N. 2010. *J. Phys. Chem. B.*, 114: 2301). All calculations are performed by Gaussian 03W suite of programs (Frisch, M. J. et al., 2004. *Gaussian 03, Revision C.02*, Wallingford, CT: Gaussian, Inc.) (Reprinted with permission from the American Chemical Society. © American Chemical Society).

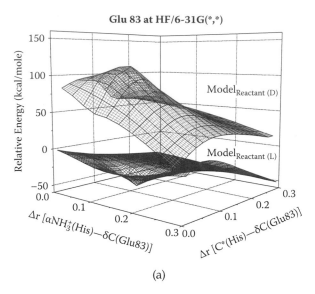

Glu 83 at HF/6-31G(*,*)

(a)

Figure 6.7 Variation in interaction energy as a function of relative distance between the nitrogen atom of the α-amino group of His and δC of Glu-83 (Å) as well as the relative distance between the chiral center of His and δC of Glu-83 (Å) for Model$_{Reactant (L)}$ and Model$_{Reactant (D)}$ with rigid geometry. Starting from the mutual arrangement of Histidine and Glu 83 as in the crystal structure, mutual separation is varied as present in the product state Model$_{Product (L)}$. Computations are performed using the two-level ONIOM (HF/6-31G**:PM3) method (a) variation in interaction energy as shown in part (a) with Glu83 at the HF/6-31G**. (Dutta Banik, S. and Nandi, N. 2010. *J. Phys. Chem. B.*, 114: 2301.) All calculations are performed by Gaussian 03W suite of programs (Frisch, M. J. et al., 2004. *Gaussian 03, Revision C.02*, Wallingford, CT: Gaussian, Inc.) (Reprinted with permission from the American Chemical Society. © American Chemical Society). (continued)

helices in the active site is needed. In all the cases studied, the reorganization is disruptive to the overall structure of the synthetase. This is another reason why the first step of the aminoacylation is highly specific for the enantiomeric form of His and the synthetase active site cavity is highly specific for L-enantiomer. The foregoing result shows that the set of active site residues confines the amino acid and ATP in a potential well using intricate interaction, which is selective of the natural L-His for adenylation.

The different functions performed by the active site residues in HisRS in *E. coli* were studied recently using the two-level ONIOM (HF/6-31G**:PM3) method (Dutta Banik and Nandi, 2010, 2301). The results of the variation of the potential energy as a function of the variation of the relative distance between the nitrogen atom of the α-amino group of His and δC of Glu83 and the chiral center of His and δC of Glu83, calculated at the HF/6-31G**:PM3 level of theory (considering Glu83 at HF level and

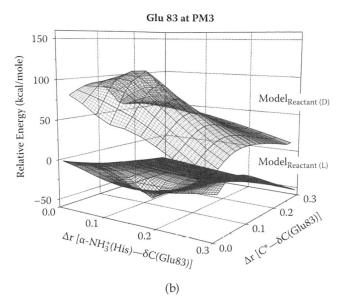

Figure 6.7 (continued) Variation in interaction energy as a function of relative distance between the nitrogen atom of the α-amino group of His and δC of Glu83 (Å) as well as the relative distance between the chiral center of His and δC of Glu83 (Å) for Model$_{Reactant (L)}$ and Model$_{Reactant (D)}$ with rigid geometry. Starting from the mutual arrangement of Histidine and Glu83 as in the crystal structure, mutual separation is varied as present in the product state Model$_{Product (L)}$. Computations are performed using two-level ONIOM (HF/6-31G**:PM3) method (b) with Glu83 considered at PM3 level. (Dutta Banik, S. and Nandi, N. 2010. *J. Phys. Chem. B.*, 114: 2301.) All calculations are performed by Gaussian 03W suite of programs (Frisch, M. J. et al., 2004. *Gaussian 03, Revision C.02*, Wallingford, CT: Gaussian, Inc.) (Reprinted with permission from the American Chemical Society. © American Chemical Society).

PM3 level, respectively) are shown in Figure 6.7a and Figure 6.7b, respectively. The plot shows that the interaction energies of the L-His and D-His with Glu83 are not identical with the variation of the distance between nitrogen atom of α-amino group of His and δC of Glu83 (indicated as Δr [α-NH$_3$$^+$ (His)···δC(Glu83)]) and with the variation of the relative distance between the chiral center of His and δC of Glu83 (indicated as Δr [C*(His)···δC(Glu83)]). The result shows that the energy surface is more favorable with increasing Δr [α-NH$_3$$^+$(His)···δC(Glu83)] (from reactant to product state), and the interaction energy of the L-His is passing through a minima, whereas that of the D-His is passing through a maxima. The difference in energy between Model$_{Reactant (L)}$ and Model$_{Reactant (D)}$ is due to the difference in the separation of the positively charged amino group of His and the negatively charged carboxylic acid group of Glu83 in the respective cases. The unfavorable nature of the energy surface of the D-His indicates that the electrostatic interaction of Glu83 with His protects against activation by the wrong enantiomer. The features of the plots at the different level of theories considered (HF or PM3) are similar, indicating that the conclusions are independent of the theoretical method used.

The variation of the energy surface as a function of relative distances between the αP atom of ATP and ξC of Arg113 (indicated as $\Delta r[\delta P(ATP)$—$\xi C(Arg113)]$) and the chiral center of His and ξC of Arg113 (indicated as $\Delta r[C^*(His)$— $\xi C(Arg113)]$) for the $Model_{Reactant\,(L)}$ as well as $Model_{Reactant\,(D)}$ calculated at the HF/.6-31G**:PM3 level of theory, including Arg 113 at HF level and PM3 level, is shown in Figure 6.8a and Figure 6.8b, respectively. The surface of $Model_{Reactant\,(D)}$ is higher in energy with respect to the energy surface of the $Model_{Reactant\,(L)}.$ The energy surface for L-His has a minimum at reduced $\Delta r[\delta P(ATP)$—$\xi C(Arg113)]$ separation and is due to the favorable electrostatic interaction between ATP and Arg113 at close separation. On the other hand, no minima is noted for $Model_{Reactant\,(D)}$ and the surface is unfavorable on average.

Figures 6.9a and 6.9b, respectively, show the variation of the potential energy as a function of variation of the relative distance between carbon atom of α carboxylic acid group of His and ξC of Arg259 (indicated as $\Delta r[\alpha COO^-(His)$—$\xi C(Arg259)]$) and chiral center of His and ξC of Arg259 (indicated as $\Delta r[C^*(His)$—$\xi C(Arg259)]$) for the $Model_{Reactant\,(L)}$ and $Model_{Reactant\,(D)}$ at the HF/6-31G **:PM3 level of theory, including Arg 259 at HF level and PM3 level, respectively. The result shows that the energy surface of $Model_{Reactant(L)}$ has a distinct minimum, which is absent in the $Model_{Reactant\,(D)}$ energy surface, and a barrier is noted in the latter case as function of the variation of the mutual separation between D-His and Arg 259. The energy is most favorable when the separation $\Delta r[\alpha COO^-$ (His)—$\xi C(Arg259)]$ is such that the interaction due to the hydrogen bonding is most favorable. The interaction is gradually lost with increase in mutual separation and is unfavorable also at the closer separation due to the short range repulsion between ATP, His, and Arg259. In contrast, the corresponding energy surface of the D-His is largely unfavorable due to the unfavorable electrostatic interaction of the positively charged amino group of His with the positively charged guanidinium group of Arg259.

The Arg259 plays an important role in discrimination as it is an important catalytic residue within the active site. The crystallographic study proposed is based on the conservation of Arg259 in the active site and its proximity to the reactants so that it might play a catalytic role. The Arg259, which is suggested to act as substitute for one of the three Mg^{2+} ions present (Mg^{2+} 1 near the α and β phosphate) (Colominas et al., 1998, 2947), is present in all class II synthetases. It is also proposed that this residue, along with Arg113, will neutralize the negative charge over the carboxyl group of His and stabilize the doubly charged pentavalent

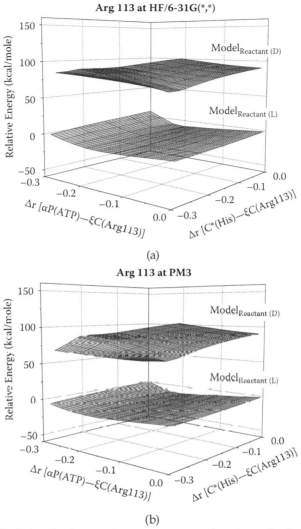

(a)

(b)

Figure 6.8 Variation in interaction energy as a function of relative distance between α phosphorous atom of ATP and ξC of Arg-113 (Å), as well as the relative distance between the chiral center of His and ξC of Arg-113 (Å) for Model$_{Reactant (L)}$ and Model$_{Reactant (D)}$ with rigid geometry. Starting from the mutual arrangement of Histidine or ATP and Arg 113 as in the crystal structure, mutual separation is varied as present in the product state Model$_{Product (L)}$. Computations are performed using two-level ONIOM (HF/6-31G**:PM3) method (a) with Arg 113 at the HF/6-31G ** (b) with Arg 113 considered at PM3 level. (Dutta Banik, S. and Nandi, N. 2010. *J. Phys. Chem. B.,* 114: 2301.) All calculations are performed by Gaussian 03W suite of programs (Frisch, M. J. et al., 2004. *Gaussian 03, Revision C.02,* Wallingford, CT: Gaussian, Inc.) (Reprinted with permission from the American Chemical Society. © American Chemical Society).

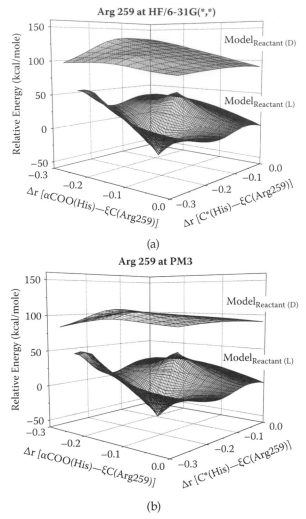

Figure 6.9 Variation in interaction energy as a function of relative distance between carbon atom of α-carboxylic acid group of His and ξC of Arg-259 (Å) as well as the relative distance between the chiral center of His and ξC of Arg-259 (Å) for Model$_{Reactant (L)}$ and Model$_{Reactant (D)}$ with rigid geometry. Starting from the mutual arrangement of Histidine and Arg 259 as in the crystal structure, mutual separation is varied as present in the product state Model$_{Product (L)}$. Computations are performed using two-level ONIOM (HF/6-31G**:PM3) method (a) with Arg 259 at the HF/6-31G** (b) with Arg 259 considered at PM3 level. (Dutta Banik, S. and Nandi, N. 2010. *J. Phys. Chem. B.*, 114: 2301.) All calculations are performed by Gaussian 03W suite of programs (Frisch, M. J. et al., 2004. *Gaussian 03, Revision C.02*, Wallingford, CT: Gaussian, Inc.) (Reprinted with permission from the American Chemical Society. © American Chemical Society).

transition state. The electronic structure-based analysis by Dutta Banik and Nandi confirms that the Arg259 plays an important catalytic role in the activation step rather than merely reducing the negative charge density over the ATP and is essentially dependent on hydrogen bonding (Dutta Banik and Nandi, 2010, 2301).

The energy surface as a function of relative distance between δN of His and ηO of Tyr 264 (indicated as $\Delta r[\delta N$ (His)—ηO(Tyr 264)]) and the chiral center of His and ηO of Tyr264 (indicated as $\Delta r[C^*$(His)—ηO(Tyr 264)]) for $Model_{Reactant\ (L)}$, as well as $Model_{Reactant\ (D)}$ with rigid geometry, is shown in the Figure 6.10a and Figure 6.10b, respectively, at the HF/6-31G**:PM3 level of theory, including Tyr264 at HF level and PM3 level, respectively. The figures show that the energies of $Model_{Reactant\ (L)}$ and $Model_{Reactant\ (D)}$ as a function of $\Delta r[\delta N$ (His)—ηO(Tyr 264)]) and $\Delta r[C^*$(His)—ηO(Tyr 264)] are more favored for the $Model_{Reactant\ (L)}$ relative to the $Model_{Reactant\ (D)}$ on an average. As the relative separation between the $\Delta r[\delta N$ (His)—ηO(Tyr 264)] decreases, the hydrogen bonding between His and Tyr264 is favored and consequently the energy surface of $Model_{Reactant(L)}$ passes through a minima. The results indicate that the interaction of L-His and the surrounding residues are more favorable than the interaction of the same residues with D-His. This shows that the network of interaction (principally electrostatic) is highly unfavorable when D-amino acid is incorporated.

The foregoing analysis of interaction of active site residues such as Glu, Arg, and Tyr indicates that the L-His is located in a network of favorable interaction, and with noncognate D-His, this interaction is highly unfavorable. The study of the variation in the energy during the mutual approach of the His and ATP to form adenylate shows that the surrounding nanospace of synthetase confines the L-His and ATP to proximally place the reactants in geometry suitable for the in-line nucleophilic attack. The confinement of reacting moieties placed within a nanometer regime of an active site drives the corresponding reaction very precisely with remarkable discrimination capacity. The significantly higher energy of the energy surface of the model containing D-His is due to the unfavorable interaction of D-His with ATP and surrounding residues. Interaction of L-His and the surrounding residues are more favorable than the interaction of the same residues with D-His. The reorganization of the surrounding nanospace can lower the unfavorable nature of the interaction energy surface of D-His and surrounding residues. However, such a rearrangement requires the placement of residues in such a manner that a large-scale structural rearrangement of the synthetase structure is needed.

The foregoing conclusion about the influence of the surrounding residues of the active site nanospace on the reaction mechanism, which is highly specific about the cognate amino acid (including its chirality), is not limited in a particular species like *E. coli* of HisRS. It is possible that the arrangement of residues in controlling the stringent chiral discrimination

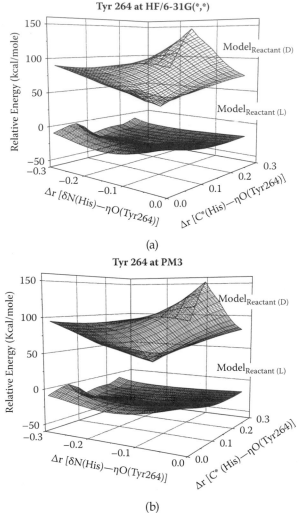

Figure 6.10 Variation in interaction energy as a function of relative distance between δN of His and ηO of Tyr-264 (Å) as well as the relative distance between the chiral center of His and ηO of Tyr-264 (Å) for Model$_{Reactant (L)}$ and Model$_{Reactant (D)}$ with rigid geometry. Starting from the mutual arrangement of Histidine and Tyr 264 as in the crystal structure mutual separation is varied as present in the product state Model$_{Product (L)}$. The Cartesian coordinates are given in Table II(c) of the supplementary material. Computations are performed using the two-level ONIOM (HF/6-31G**:PM3) method (a) with Tyr 264 at the HF/6-31G** (b) with Tyr 264 considered at PM3 level. (Dutta Banik, S. and Nandi, N. 2010. *J. Phys. Chem. B.*, 114: 2301.) All calculations are performed by Gaussian 03W suite of programs. (Frisch, M. J. et al., 2004. *Gaussian 03, Revision C.02*, Wallingford, CT: Gaussian, Inc.) (Reprinted with permission from the American Chemical Society. © American Chemical Society).

is developed through evolution and retained thereafter. Although the dissimilarity in the amino acid binding pocket is observed for different aaRSs, the architecture of the active site of a particular aaRS from a different organism is analogous. Crystallographic and biochemical studies revealed that a number of active site residues are conserved in different species for a given aaRS (Ibba and Söll, 2000, 617; Sankaranarayanan and Moras, 2001, 323; Arnez et al., 1995, 4143; Calendar and Berg, 1966, 1690; Soutourina et al., 2000, 32535; Bergmann et al., 1961, 1735; Norton et al., 1963, 269; Yamane et al., 1974, 4135; Tamura and Schimmel, 2004, 1253; Francklyn et al., 2008, 100). Enzyme superfamilies exhibit remarkable variations regarding the substrate specificity and reaction chemistry as well as concerning the presence of the catalytic residues in the active site (Todd et al., 2001, 1113). Computational studies for a few aaRSs have revealed diverse roles of these residues such as recognition, catalysis, chiral specificity, and substrate positioning (Archontis et al., 1998, 823; Dutta Banik and Nandi, 2009b, 468; Nandi, 2009, 111; Dutta Banik and Nandi, 2010, 2301; Thompson et al., 2007, 30856). Essentially, these active site residues and their architectonics control the mechanism of the reaction, which is remarkably accurate. This accuracy is developed through evolution to make the process of protein synthesis error-free. The active site structure, so perfected by evolution, is possibly retained further. Dutta Banik and Nandi compared the primary sequence and architecture, as well as the reaction mechanism of the first step of aminoacylation reaction in three different species of histidyl-tRNA synthetase (HisRS), using a quantum chemical method (Dutta Banik and Nandi, 2011, in press).

A comparison of the primary sequences of HisRS of *E. coli* (Arnez et al., 1997, 714427), *TT* (Åberg, 1997, 3084, 28), and *SA* (Qiu et al., 1999, 12296), and the reaction mechanisms in these species was made recently (Dutta Banik and Nandi, 2011, in press). Despite the gaps in the primary sequence, the corresponding higher-level structure is so organized that the conserved active site residues are placed in specific locations relative to the reactants. These specific conserved residues (such as Glu83, Arg113, Gln127, Glu130, Arg259, Tyr264, Gly304, Phe305, Ala306, etc., as in *E. coli*) in the active site are located in a three-dimensional arrangement, which have close similarity in the species concerned (Dutta Banik and Nandi, 2011, in press). This indicates that the organization of the secondary and tertiary structural units of three enzymes is such that the nanodimensions of the active site of the three species bear close resemblance, despite the gaps in the sequences.

The sequence alignment of HisRS from three different organisms shows a single and double residue gap in between SI loop and motif 2 loop for *E. coli* and *TT*, respectively. However, the overall features of the three active sites are qualitatively similar based on the relative arrangement of secondary structural elements as shown in Figure 6.11. Despite the gaps in sequences, the relative separations between the active site

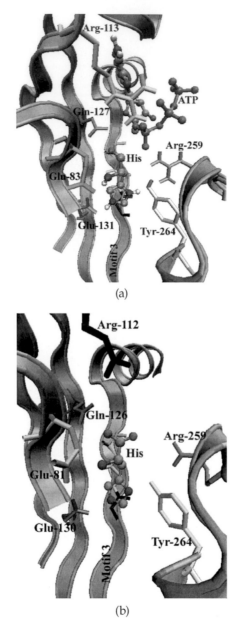

(a)

(b)

Figure 6.11 **See color insert**. (a) The representation of the active site of three pro-karyotic HisRS *E. coli*, (b) *TT*, and (c) *SA* as obtained from the respective crystal structure. The images are prepared using VMD (Humphrey et al., 1996: 33). (continued)

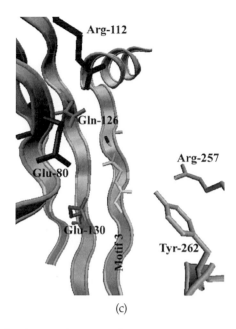

(c)

Figure 6.11 **See color insert.** (continued) (a) The representation of the active site of three prokaryotic HisRS *E. coli*, (b) *TT*, and (c) *SA* as obtained from the respective crystal structure. The images are prepared using VMD (Humphrey et al., 1996: 33).

residues present in these loops (as measured between the chiral centers of the residues) Glu83 (Glu81, Glu80) and Arg113 (Arg112, Arg112) belonging to the motif 2 loop, are nearly the same. This indicates that the organization of the secondary structural elements of three enzymes is such that the nanodimension of the active site of the three species are closely similar, despite the differences in the sequences.

However, the spatial organizations of the residues comprising the walls of the active site display both similarity and dissimilarity in the three species of HisRS. While a number of residues are common, others are not, as noted in Figure 6.12. Among the conserved residues, the Arg259 (Arg259, Arg257) and Arg113 (Arg112, Arg112) have a known catalytic role, while the Glu83 (Glu81, Glu80) and Tyr264 (Tyr264, Tyr262) have roles in controlling the fidelity of the reaction. The nonpolar residues such as Ala, Phe, and Gly are also conserved. These residues are in close proximity to the hydrophobic part of the side chain of the reactant amino acid (His), rather than its charged terminals. Hence, those residues are conserved that have a catalytic role (Arg259), have a role in reducing the unfavorable electrostatic charge and helping in the progress of reaction (Arg113), control the fidelity of the reaction (Glu 83 or Tyr 264), or have influence in placing the reactant in a suitable orientation by hydrophobic

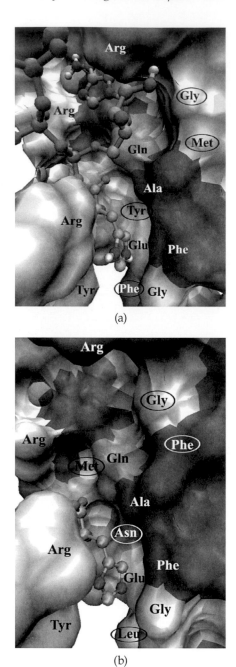

Figure 6.12 **See color insert**. Comparison of the active site residues in HisRS of three prokaryotic species, *E. coli*, *TT*, and *SA*. The residues which are nonconserved in three prokaryotic HisRS (a) *E. coli* and (b) *TT* are encircled in the figure. The images are prepared using VMD (Humphrey et al., 1996: 33). (continued)

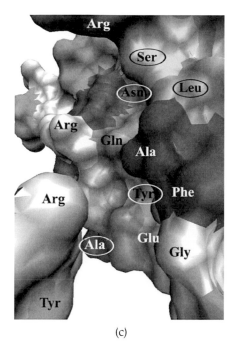

(c)

Figure 6.12 **See color insert**. (continued) Comparison of the active site residues in HisRS of three prokaryotic species, *E. coli, TT,* and *SA.* The residues which are nonconserved in three prokaryotic HisRS (c) *SA* are encircled in the figure. The images are prepared using VMD (Humphrey et al., 1996: 33).

interaction (Ala-Phe-Gly triad). Other residues like Tyr107, Phe133, Met307, and Gly308 in *E. coli*, Asn128, Leu132, Phe307, and Gly308 in *TT* and Ala83, Tyr106, Asn125, Leu308, and Ser309 in *SA* differ in one or more species considered here, although they comprise the walls of the active site.

An electronic structure-based analysis was necessary to understand the structure-based hypothesis that a conserved residue has the particular role as substrate binding (ionic interaction, hydrogen bonding, and hydrophobic interaction), charge neutralization, and catalytic action in the active site of all corresponding species where the presence of such residue is retained by evolution. Electronic structure-based methods are useful in studying the corresponding potential energy surfaces, transition state barrier height, and stabilizing interactions of the transition state. A transition state is the first-order saddle point where the energy passes through a maximum along the pathway that connects the minima, but is a minimum for displacements in all other directions perpendicular to the path. This saddle point is characterized by one negative value of the corresponding Hessian matrix (referred to as the "imaginary frequency") (Leach, 2001, 279). Hence, a transition state can be identified by the frequency analysis and cannot be confirmed unambiguously based on the information about the thermodynamically average states as obtained from crystal structure analysis. The transition state structure analysis in HisRS points out another factor for chiral discrimination that the transition state geometry containing D-His lacks the hydrogen bond network as present in the case of L-His, which makes the reaction path unfavorable (Dutta Banik and Nandi, 2010, 2301).

The computational studies (Dutta Banik and Nandi, 2011, in press; 2010, 2301) confirm that the aminoacylation reaction follows a common reaction pathway for the three species and is dependent on the conserved active site residues in close proximity to the reactants. The crystal structures of the oligomeric complex of histidyl-tRNA synthetase (HisRS) from *E. coli* complexed with ATP and histidinol and histidyl-adenylate (1KMM. PDB and 1KMN.PDB) (Arnez et al., 1997, 7144) are used to generate the models of active site of reactant and products used in theoretical calculation (Dutta Banik and Nandi, 2010, 2301). The model of the active site also includes Glu83, Arg113, Gln127, Glu131, Arg259, Tyr264, Gly304, Phy305, and Ala306. These residues belong to different motifs or loops and are in close proximity to the reactants in the active site. The model also includes two Mg^{2+} ions located near the β-γ phosphate linkage. As mentioned before, the Arg 259, which is suggested to act as substitute for one of the three Mg^{2+} ions present (Mg^{2+} 1 near the α and β phosphate) (Colominas et al., 1998, 2947), as present in all class II synthetases is also included. Two water molecules are included. The structure is optimized using two-level ONIOM (HF/6-31G**:PM3) method (Dutta Banik and Nandi, 2010, 2301). The histidine, and α and β phosphorus of ATP with attached oxygen and the sugar ring, are considered as high level and all the surrounding residues are considered as low level. The electrostatic potential for reactants (His and ATP), as well as surrounding residues of active site are computed using 6-31G** basis set using Mulliken population analysis. The removal of the conserved active site residues and concomitant enhancement of barrier height indicates that the conserved active site residues cooperatively lower the reaction rate by several orders. This result shows that the active site cooperatively (with its constituent amino acids, which are close to the reactants) reduces the barrier height of the reaction. These residues form a network of electrostatic interaction that are favorable for cognate but unfavorable for the enantiomeric form or other noncognate forms (such as with a different charge over the side chain). The study shows that a conserved structural pattern is responsible for the aminoacylation reaction.

An analysis of the transition states and the interaction energy profiles confirms that the mechanisms are closely similar, despite the differences in the active sites of different species. The conserved active site residues cooperatively lower the reaction rate by several orders and help to retain the fidelity of reaction. A comparison of the optimized geometries of the respective transition states of *E. coli*, *TT*, and *SA* confirms that the reaction proceeds through a penta-coordinated transition state for three organisms. The potential energy surface of the active site in the reactant state is complex as the reaction coordinate space is large for such biological systems. In general, such a complex system is expected to have a multitude of minima (Wales et al., 1998, 758). A comparison of the variation in interaction energies as a function of the identified variables for the incorporation of L-amino acid and D-amino acid is expected to reveal the fidelity mechanism. Such variables are inter- as well as intramolecular coordinates of reactant (His, ATP) and surrounding active site residues. These variables significantly change during the reaction. The relative difference in the transition state barrier height for the reaction between L- or D-His and ATP is also expected to be different and contribute to the discrimination, which can contribute to the chiral specificity in the activation step of aminoacylation.

Preliminary studies support the foregoing view (Dutta Banik and Nandi, 2009b, 468). In such complex systems, individual minima are clustered into "basins" separated by energy barriers, and the internal distribution between basins is separated by a small energy barrier (close to $k_B T$) (Becker and Karplus, 1997, 1495). As mentioned before, a penta coordinated transition state for the activation reaction is proposed from crystallographic studies. Such pentacovalent phosphorous intermediates or transition states formed by nucleophilic attack on the tetracoordinated phosphorous atom have biological significance, and have been analyzed by *ab initio* studies (Lim and Karplus, 1990, 5872). Hence, an electronic structure-based analysis is expected to be useful in understanding the reaction mechanism.

The energy profiles corresponding to the progress of the reaction are studied in three species as a function of two variables that describe the approach of the reactants toward each other and the effect of the influence of the major catalytic residue, Arg 259. The first one is given by $q = r_{P-O(PPi)} - (r_{P-O(His)} + r')$, where the subscripts indicate the bond. The r' is the projection of the $\alpha P\text{-}O_5'$ bond of ATP on the bond forming between the carboxylic oxygen atom of His and the α-phosphorus atom of ATP. While the magnitude of the $r_{P-O(PPi)}$ gradually increases, the magnitude of the r' decreases, and as a result q increases during the progress of the reaction toward the product state. The second variable measures the progress of the reaction as a function of the orientation of the catalytic residue Arg259. The change in orientation of this residue is important in successfully converting the reactant state to the product one. The corresponding contour diagram as a function of q and the orientation of the Arg259 are shown in Figure 6.13. The progress of the aminoacylation reaction follows a common reaction pathway. This is critically dependent on the conserved active site residues in close proximity to the reactants, each of which performs a specific role. The roles, although varying from catalyzing the

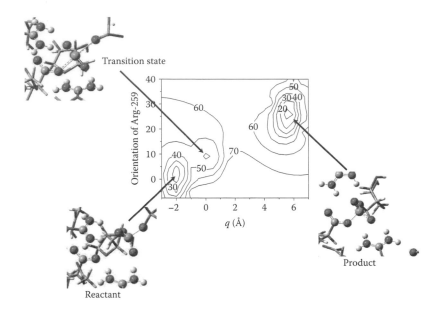

Figure 6.13 The contour representation of the variation in energy during the progress of the activation step of aminoacylation reaction in the model of active site of HisRS of *E. coli* (Dutta Banik, S. and Nandi, N. 2011, in press. The variation of interaction energy as a function of the reaction coordinate (q) and orientation of the Arg-259 (Arg259, Arg257) is calculate at ONIOM HF/6-31G*:PM3//HF/6-31G*:PM3 level. The relative orientation of the Arg259 with respect to the His and ATP as present in the reactant state is taken as initial orientation (referred to as zero degrees). Part of the calculation is carried out using Gaussian 03W suite of programs (Frisch, M. J. et al., 2004. *Gaussian 03, Revision C.02*, Wallingford, CT: Gaussian, Inc.). The representative optimized geometries of the reactant, product, and transition state are indicated in the diagram.

reaction, reduce the unfavorable electrostatic charges to the positioning of substrate in a suitable orientation for reaction, and are common in the three species (Dutta Banik and Nandi, 2010, 2301).

    The computational studies in aaRS indicate that the capacity to exhibit stringent chiral discrimination is critically dependent on the conserved active site residues in close proximity to the reactants, each of which performs a specific role. The roles, although varying from catalyzing the reaction and reducing the unfavorable electrostatic charges to facilitate the reaction to the task of positioning of substrate in a suitable orientation for reaction, are common in different species. The computational studies (Dutta Banik and Nandi, 2009b, 468; Nandi, 2009, 111; Dutta Banik and Nandi, 2011, in press; 2010, 2301) quantitatively show for the first time the specific conserved structural patterns as used to perform the aminoacylation reaction in HisRS. Although the Arg259 plays a significant catalytic role, other conserved residues also influence the nature of the transition state. A comparison of the transition state barrier heights in the presence and absence of different active site surrounding residues of HisRS of *E. coli* shows that the barrier height abruptly increases as the conserved active site residues are removed from the model. This indicates a very large difference in the rates of the aminoacylation reaction in the presence and the absence of the active site residues. The result clearly indicates that the conserved active site residues (such as Glu83, Arg113, Gln127, and Arg259) play a more important role rather than merely anchoring the reactants at a suitable orientation in the reaction and are responsible for the stabilization of transition state, as well as retention of enantiopurity. While a few residues such as Arg259 have a dominant role such as catalytic activity, the active site as a whole (with its constituent amino acids, which are close to the reactants) helps reduce the barrier height, compared to the reaction in the absence of the active site.

References

Åberg, A., Yaremchuk, A., Tukalo, M., Rasmussen, B., and Cusack, S. (1997). Crystal structure analysis of the activation of histidine by *Thermus thermophilus* Histidyl-tRNA Synthetase. *Biochemistry*, 36: 3084–3094.

Archontis, G., Simonson, T., Moras, D., and Karplus, M. (1998). Specific amino acid recognition by aspartyl-tRNA synthetase studied by free energy simulations *J. Mol. Biol.*, 275: 823–846.

Arnez, J.G., Dock-Bregeon, A., and Moras, D. (1999). Glycyl-tRNA synthetase uses a negatively charged pit for specific recognition and activation of glycine. *J. Mol. Biol.*, 286: 1449–1459.

Arnez, J.G., Augstine, J.G., Moras, D., and Francklyn, C.S. (1997). The first step of aminoacylation at the atomic level in histidyl-tRNA synthetase. *Proc. Natl. Acad. Sci. (USA)*, 94: 7144–7149.

Arnez, J.G., Harris, D.C., Mitschler, A., Rees, B., Francklyn, C.S., and Moras, D. (1995). Crystal structure of histidyl-tRNA synthetase from *Escherichia coli* complexed with histidyl-adenylate. *EMBO J.*, 14: 4143–4155.

Becker, O.M. and Karplus, M. (1997). The topology of multidimensional potential energy surfaces: Theory and application to peptide structure and kinetics. *J. Chem. Phys.*, 106: 1495–1517.

Belrhali, H., Yaremchuk, A., Tukalo, M., Berthet-Colominas, C., Rasmussen, B., Bosecke, P., Diat, O., and Cusack, S. (1995). The structural basis for seryladenylate and Ap_4A synthesis by seryl-tRNA synthetase. *Structure*, 3: 341–352.

Benedicte, D., Moras, D., and Cavarelli, J. (2000). tRNA aminoacylation by arginyl-tRNA synthetase: Induced conformations during substrates binding. *EMBO J.*, 19: 5599–5610.

Berg, J.M., Tymoczko, J.L., and Stryer, L. (2002). *Biochemistry*, New York: W.H. Freeman, 671–672.

Bergmann, F., Berg P., and Dieckmann, M. (1961). The enzymic synthesis of amino acyl derivatives of ribonucleic acid. *J. Biol. Chem.*, 236: 1735–1740.

Biou, V., Yaremchuk, A., Tukalo, M., and Cusack, S. (1994). The 2.9 Å crystal structure of *T. thermophilus* seryl-tRNA synthetase complexed with tRNA[Ser]. *Science*, 263: 1404–1410.

Calendar, R. and Berg, P. (1967). D-tyrosyl RNA: Formation, hydrolysis and utilization for protein synthesis. *J. Mol. Biol.*, 26: 39–54.

Calendar, R. and Berg, P. (1966). The catalytic properties of tyrosyl ribonucleic acid synthetases from *Escherichia coli* and *Bacillus subtilis*. *Biochemistry*, 5: 1690–1695.

Cavarelli, J., Delagoutte, B., Eriani, G., Gangloff, J., and Moras, D. (1998). L-arginine recognition by yeast arginyl tRNA synthetase. *EMBO J.*, 17: 5438–5448.

Cavarelli, J., Eriani, G., Rees, B., Ruff, M., Boeglin, M., Mitschler, A., Martin, F., Gangloff, J., Thierry, J.C., and Moras, D. (1994). The active site of yeast aspartyl-tRNA synthetase: Structural and functional aspects of the aminoacylation reaction. *EMBO J.*, 13: 327–337.

Colominas, C., Seignovert, L., Härtlein, M., Grotli, M., Cusack, S., and Leberman, R. (1998). The crystal structure of asparaginyl-tRNA synthetase from *Thermus thermophilus and* its complexes with ATP and asparaginyl-adenylate: The mechanism of discrimination between asparagines and aspartic acid. *EMBO J.*, 17: 2947–2960.

Cusack, S., Yaremchuk, A., and Tukalo, M. (2000). The 2Å crystal structure of leucyl-tRNA synthetase and its complex with a leucyl-adenylate analogue. *EMBO J.*, 19: 2351–2361.

Davie, E., Konigsberger V., and Lipman, F. (1956). The isolation of a tryptophan-activating enzyme from pancreas. *Arch. Biochem. Biophys.*, 65: 21–38.

Desogus, G., Todone, F., Brick P., and Onesti, S. (2000). Active site of lysyl-tRNA synthetase: Structural studies of the adenylation reaction. *Biochemistry*, 39: 8418–8425.

Dutta Banik, S. and Nandi, N. (2009a). Molecular understanding of the influence of chirality and interaction in soft systems: From biomimetics to biosystems, a retrospection. *J. Surf. Sci. Technol.*, 25: 1–18.

Dutta Banik, S. and Nandi, N. (2009b). Orientation and distance dependent chiral discrimination in the first step of the aminoacylation reaction: Integrated molecular orbital and semi-empirical method (ONIOM) based calculation. *Coll. Surf. B: Biointerfaces*, 74: 468–476.

Dutta Banik, S. and Nandi, N. (2010). Aminoacylation reaction in the histidyl-tRNA synthetase: Fidelity mechanism of the activation step. *J. Phys. Chem. B.*, 114: 2301–2311.

Dutta Banik, S. and Nandi, N. (2011). Influence of the conserved active sites residues of Histidyl tRNA synthetase on the mechanism of aminoacylation reaction. *Biophys. Chem.* (in press).

Eiler, S., Dock-Bregeon, A.C., Moulinier, L., Thierry, J.C., and Moras, D. (1999). Synthesis of aspartyl-tRNA[Asp] in *Escherichia coli*: A snapshot of the second step. *EMBO J.*, 18: 6532–6541.

Fersht A.R. and Dingwall, C. (1979). Cysteinyl-tRNA synthetase from *Escherichia coli* does not need an editing mechanism to reject serine and alanine. High binding energy of small groups in specific molecular interactions. *Biochemistry*, 18: 1245–1249.

Fersht, A.R. (1986). *Accuracy in Molecular Processes*, Kirkwood, T.B.L. Rosenberg, R.F. Galas, D.J. (eds), Chapman & Hall, New York, 69–82.

Fersht, A.R. (1987). Dissection of the structure and activity of the tyrosyl-tRNA synthetase by site-directed mutagenesis. *Biochemistry*, 26: 8031–8037.

Francklyn, C.S., First, E.A., Perona, J.J., and Hou, Y. (2008). Methods for kinetic and thermodynamic analysis of aminoacyl tRNA synthetases. *Methods*, 44: 100–118.

Fersht, A.R., Knill-Jones, J.W., Bedouelle H., and Winter, G. (1988). Reconstruction by site-directed mutagenesis of the transition state for the activation of tyrosine by the tyrosyl-tRNA synthetase: A mobile loop envelopes the transition state in an induced-fit mechanism. *Biochemistry*, 27: 1581–1587.

Frisch, M.J. et al. (2004). *Gaussian 03, Revision C.02*, Wallingford, CT: Gaussian, Inc.

Fukai, S., Nureki, O., Sekine, S., Atsushi Shimada, Tao, J., Vassylyev, D.G., and Yokoyama, S. (2000). Structural basis for double-sieve discrimination of L-valine from L-isoleucine and L-threonine by the complex of tRNA^{Val} and valyl-tRNA synthetase. *Cell*, 103: 793–803.

Hecht, S. (1992). Probing the synthetic capabilities of a center of biochemical catalysis. *Acc. Chem. Res.*, 25: 545–552.

Hopfield, J.J. (1974). Kinetic proofreading: A new mechanism for reducing errors in biosynthetic processes requiring high specificity. *Proc. Natl. Acad. Sci. (USA)*, 71: 4135–4139.

Humphrey, W., Dalke, A., and Schulten, K. (1996). VMD: Visual molecular dynamics. *J. Mol. Graphics*, 114: 33–38.

Ibba, M. and Söll, D. (2000). Aminoacyl-tRNA synthesis. *Annu. Rev. Biochem.*, 69: 617–650.

Leach, A.R. (2001). *Molecular Modelling: Principles and Applications*, 2nd ed. Pearson, Prentice Hall, 279.

Lim, C. and Karplus, M. (1990). Nonexistence of dianionic pentacovalent intermediates in an *ab initio* study of the base-catalyzed hydrolysis of ethylene phosphate. *J. Am. Chem. Soc.*, 112: 5872–5873.

Nakama, T., Nureki, O., and Yokoyama, S. (2001). Structural basis for the recognition of isoleucyl-adenylate and an antibiotic, mupirocin, by isoleucyl-tRNA synthetase. *J Biol. Chem.*, 276: 47387–47393.

Nandi N. (2005). Study of chiral recognition of model peptides and odorants: Carvone and camphor. *Curr. Sci.*, 88: 1929–1937.

Nandi, N. (2003). Molecular origin of the recognition of chiral odorant by chiral lipid: Interaction of dipalmitoyl phosphatidyl choline and carvone. *J. Phys. Chem. A.*, 107: 4588–4591.

Nandi, N. (2009). Chiral discrimination in the confined environment of biological nanospace: Reactions and interactions involving amino acids and peptides. *Int. Rev. Phys. Chem.*, 28: 111–167.

Newberry, K.J., Hou, Y.-M., and Perona, J.J. (2002). Structural origins of amino acid selection without editing by cysteinyl-tRNA synthetase. *EMBO J.*, 21: 2778–2787.

Norton, S., Ravel, J., Lee C., and Shive, W. (1963). Purification and properties of the aspartyl ribonucleic acid synthetase of *Lactobacillus arabinosus*. *J. Biol. Chem.*, 238: 269–274.

Onesti, S., Desogus, G., Brevet, A., Chen, J., Plateau, P., Blanquet, S., and P. Brick, (2000). Structural studies of lysyl-tRNA synthetase: Conformational changes induced by substrate binding. *Biochemistry*, 39: 12853–12861.

Perona, J.J., Rould, M.A., and Steitz, T.A. (1993). Structural basis for transfer RNA aminoacylation by *Escherichia coli* glutaminyl-tRNA synthetase. *Biochemistry*, 32: 8758–8771.

Profy, A.T. and Usher, D.A. (1984). Stereoselective aminoacylation of a dinucleoside monophosphate by the imidazolides of DL-alanine and N-(*tert*-butoxycarbonyl)-DL-alanine. *J. Mol. Evol.*, 20: 147–156.

Prosser, I., Altug, I.G., Phillips, A.L., König, W.A., Bouwmeester, H.J., and Beale, M.H. (2004). Enantiospecific (+)- and (-)-germacrene D synthases, cloned from goldenrod, reveal a functionally active variant of the universal isoprenoid-biosynthesis aspartate-rich motif. *Archives Biochem. Biophys.*, 432: 136–144.

Qiu, X., Janson, C.A., Blackburn, M.N., Chhohan, I.K., Hibbs, M., and Abdel-Meguid, S.S. (1999). Cooperative structural dynamics and a novel fidelity mechanism in histidyl-tRNA synthetases. *Biochemistry*, 38: 12296–12304.

Rath, V.L., Silvian, L.F., Beijer, B., Sproat, B.S., and Steitz, T.A. (1998). How glutaminyl-tRNA synthetase selects glutamine. *Structure*, 6: 439–449.

Sankaranarayanan R. and Moras, D. (2001). The fidelity of the translation of the genetic code. *Acta Biochim. Polonica*, 48: 323–335.

Sekine, S., Nureki, O., Dubois, D.Y., Bernier, S., Chenevert, R., Lapointe, J., Vassylyev, D.G., and Yokoyama, S. (2003). ATP binding by glutaminyl-tRNA synthetase is switched to the productive mode by tRNA binding. *EMBO J.*, 22: 676–688.

Soutourina, J., Plateau P., and Blanquet, S. (2000). Metabolism of D-aminoacyl-tRNAs in *Escherichia coli* and *Saccharomyces cerevisiae* cells. *J. Biol. Chem.*, 275: 32535–32542.

Stranden, M., Borg-Karlson, A.K., and Mustaparta, H. (2004). Receptor neuron discrimination of the germacrene D enantiomers in the moth *Helicoverpa armigera*. *Chem. Senses.*, 27: 143–152.

Tamura, K. and Schimmel, P. (2004). Chiral-selective aminoacylation of an RNA minihelix. *Science*, 305: 1253.

Tamura, K. and Schimmel, P. (2006). Chiral-selective aminoacylation of an RNA minihelix: Mechanistic features and chiral suppression. *Proc. Natl. Acad. Sci. (USA)*, 103: 13750–13752.

Thompson, D., Lazennec, C., Plateau P., and Simonson, T. (2007). Ammonium scanning in an enzyme active site the chiral specificity of aspartyl-tRNA synthetase. *J. Biol. Chem.*, 282: 30856–30868.

Todd, A.E., Orengo, C.A., and Thornton, J.M. (2001). Evolution of function in protein superfamilies from a structural perspective. *J. Mol. Biol.*, 307: 1113–1143.

Wales, D.J., Miller, M.A., and Walsh, T.R. (1998). Archetypal energy landscapes. *Nature*, 394: 758–760.

Warshel, A., Sharma, P.K., Kato, M., Xiang, Y., Liu, H., and Olsson, M.H.M. (2006). Electrostatic basis for enzyme catalysis. *Chem. Rev.*, 106: 3210–3235.

Xin, Y., Li, W., and First, E.A. (2000). Stabilization of the transition state for the transfer of tyrosine to $tRNA^{Tyr}$ by tyrosyl-tRNA synthetase. *J. Mol. Biol.*, 303: 299–310.

Yamane, T. and Hopfield, J.J. (1977). Experimental evidence for kinetic proofreading in the aminoacylation of tRNA by synthetase. *Proc. Natl. Acad. Sci. (USA)*, 74: 2246–2250.

Yamane, T., Miller, L., and Hopfield, J.J. (1974). Kinetic proofreading: A new mechanism for reducing errors in biosynthetic processes requiring high specificity. *Proc. Natl. Acad. Sci. (USA)*, 71: 4135–4139.

Yamasaki, T., Sato, M., and Sakoguchi, H. (1997). (-)-Germacrene D: Masking substance of attractants for the cerambycid beetle, *Monochamus alternatus* (Hope). *Appl. Entomol. Zool.*, 32: 423–429.

Yaremchuk, A., Kriklivyi, I., Tukalo, M., and Cusack, S. (2002). Class I tyrosyl-tRNA synthetase has a class II mode of cognate tRNA recognition. *EMBO J.*, 21: 3829–3840.

Yaremchuk, A., Tukalo, M., Grøtli, M., and Cusack, S. (2001). A succession of substrate induced conformational changes ensures the amino acid specificity of *Thermus thermophilus* prolyl-tRNA synthetase: Comparison with histidyl-tRNA synthetase. *J. Mol. Biol.*, 309: 989–1002.

chapter seven

Summary and future directions

In the Introduction of this book, questions are posed as to the correlation between the omnipresence of chirality in naturally occurring molecules and their functionality. Enzymes are molecular machines having stringent chiral discrimination within their active sites. Analysis of the molecular organization of the active site structure and of the related interactions is useful to understand the chiral specificity of enzymatic reactions. Crystallographic analysis provides clear insight into the microscopic organization of the enzymatic structures. Electronic structure-based computations and molecular dynamics studies provide complementary views about enzymatic reactions. The wealth of information now forms a platform for analysis of the origin of the chiral preference exhibited by various enzymes. Active sites are nanoflasks where such reactions are occurring, and it is instructive to look into their molecular processes so that one can learn about the generic mechanisms used by these natural devices to utilize the cognate substrate and reject the others in a fast and accurate way. Although different enzymes, and even enzymes in a given class, can catalyze reactions of a bewildering variety, studies so far have been greatly promising in understanding the role of chirality in biological reactions and in following the fundamental correlation between chirality and life. Notably, the mechanism of discrimination of one substrate from the other is key to the ability of the enzymes to act as efficient biocatalysts, which is hard to match by synthetic design and which provides impetus for the study of enzymatic biocatalysis.

This book presents a perspective on the microscopic understanding of the influence of the active sites of different classes of enzymes on the process of chiral discrimination in the course of biological reactions. It aims to gather current understanding about the interactions present in the active site of enzymes that influence chiral discrimination. Examples are discussed where the interactions between the active site residues and the reacting species are either known or are understood. The concomitant influences on the discrimination, chiral specificity, and fidelity of the reaction are studied by experiment or computational techniques or both. The related studies indicate that the conserved residues might carry out specific roles in the reaction. Both microscopic knowledge of the detailed structure of the active sites, as well as the electronic structure-based analysis of the network of interaction, are necessary for understanding the specific roles. A dearth of knowledge about the suitable crystal structure or lacunae in the network of interaction can make understanding chiral discrimination difficult in some cases. However, where such information is available, the results are promising.

It is to be noted that isomerase and epimerase are the classes of enzymes that have the capacity to carry out the inversion of stereochemistry. Isomerase and epimerase enzymes are different from the others mentioned before in that they have the capacity to stabilize two stereoisomeric transition states and are evolved to be deficient in stereospecificity (Tanner, 2002, 237). Since the isomerase and epimerase enzymes perform the process of racemization, which is an act somewhat opposite to chiral discrimination, the influence of the active site of isomerase or epimerase and their influence on the related reaction are not discussed.

Following the foregoing perspective, various chapters of this book aim to bring together the active site structure, chiral discriminations exhibited, and correlations between the two in different classes of enzymes. This book starts with a brief introduction of chirality and chiral discrimination. A few questions pertaining to chiral discrimination in biological systems have been raised, namely, about the influence of the chirality of the amino acids and sugars themselves and their mutual influence on enzymatic reactions. Discussions of chiral discrimination in biomimetic systems are introduced. These are followed by a brief introduction to chiral discrimination in biological systems and reactions in the active sites of enzymes. The active site of an enzyme is a nanodimensional space that confines reactants and places them in a suitable orientation for reaction to happen. The effect of the confinement of the reactants on chiral discrimination is noted to be significant.

The organization of the active site of cytochrome P450, alcohol oxidoreductase, LOX, COX, nitric oxide synthase, and enoyl reductase and interactions therein are discussed in Chapter 2. The discrimination exhibited by peptidyl transferase, telomerase, HIV-1 reverse transcriptase, and DNA polymerase is discussed in Chapter 3. In the case of transferases, it has been observed that during peptide synthesis in peptidyl transferase it is impossible to incorporate D-amino acid into proteins without modifying the chemical structure of the PTC (peptidyl transferase center) of naturally occurring tRNA. However, it is possible to perform the formation of the D-L peptide bond in the model or modified systems. Computational studies show that the chiral discrimination is not only driven by the chirality of the A- and P-terminals, but it is also due to the influence of the chirality of the D-sugar ring as well as the confinement by the surrounding bases of the PTC. The nanoscale confinement of these terminals is important for the discrimination, and at larger separation, discrimination is lost. The restricted nature of the mutual orientation between the terminals during the rotatory path for the approach to form the peptide bond makes the resultant interaction profile for L-L and D-L pair different. The D-sugar ring favorably influences the rotatory path for the approach of the A- and P-terminals to form the peptide bond. The interaction of the D-sugar with the l-amino acids is more favorable than other homo- or heteropair combinations of sugar–amino acid chiralities. The stereochemistry at the 2' center of the D-sugar is vital, as it catalyzes peptide bond formation by making proper placement of the OH group that is involved in the catalysis. The nanoscale proximity of some of the surrounding bases present in PTC with the A- and P-terminals and their restricted orientation influences the discrimination. Chapter 4 discusses the chiral discrimination by epoxide hydrolases and lipases where the discrimination can be understood from the active site structure of the respective enzymes, considering their spatial arrangement and concomitant favorable interaction with the cognate substrate. In Chapter 5, chiral discrimination in lyases such as hydroxynitrile lyase and chondroitin lyases is discussed.

Chapter 6 discusses the discrimination in ligases such as germacrene D synthase and aminoacyl tRNA synthetase. The aaRS is a ligase where the influence of the active site on the reaction is studied in detail. The aminoacylation reaction is strongly dependent on the chirality of the substrate and active site residues and their mutual position and orientation. In the case of aaRS, the chiralities of the substrates of the active site amino acids, their confinement, and proximity in the cavity in a nanoscale range, and the resultant network of electrostatic interaction is noted to play a significant role. The charge distribution in and around the active site plays an important role. The restricted nature of the mutual orientation within the cavity of the active site makes the resultant interaction profile different for L- and D-amino acid exhibiting discrimination. Alteration of the chirality of the amino acid destabilizes the transition state as well as removes favorable electrostatic interaction between the surrounding residue present in the active site and the amino acid substrate.

Various interactions between the active site residues of a given aaRS and its cognate amino acid can be categorized as substrate binding (ionic interaction, hydrogen bonding, and hydrophobic interaction), charge neutralization, and catalytic action. These functional roles are expected to be played by one or more active site residues of each of the different aaRSs during the course of the reaction. These functional residues are conserved in the active site and are distributed through the regions of the active site surrounding the substrate moieties (ATP binding pocket, the amino acid binding pocket, and the reaction center in the case of aaRS) on the basis of their function. The interaction patterns with substrates and active site residues bear commonalities in different species. Presumably, the residues forming such common interaction patterns for an amino acid are conserved through evolution in different species.

Such an understanding emerges from the combination of structural analysis and computational understanding. It is worthwhile to explore the unifying principles underlying the preferential interaction and discrimination as exhibited by various enzymes in general. One obvious utility is that it might be possible to design models of active sites that can perform reactions involving new reactants with high fidelity using amino acids or nucleotides. Unless such an approach supplements the experimental approaches for designing biocatalysts, the attempt will remain random trial-and-error rather than logic-based design. Since the enzyme engineering is dependent on the generation of variants of template enzymes and identification of best-suited candidates, quantitative understanding of interactions is ideal for effective design.

Although enzymes are optimized by evolution, their structures might undergo adaptive evolution following the influence of chemicals or drugs (Marti et al., 2008, 2634). An enzyme structure might evolve over a certain period of time and adopt new enzymatic functions. It has been pointed out that the active sites of the enzymes can be used as templates to develop new enzymes that can catalyze the formation of novel materials. Computational analysis of the interactions of the active site and the reactants provides an opportunity to identify the origin of the catalytic activity and specificity using a quantitative rather than trial-and-error approach. Structure-based methods (application of classical or quantum mechanical methods to models of active sites) and reaction-analysis-based methods (application of hybrid quantum mechanical/molecular mechanical methods to follow the chemical reaction using the dynamical system) are used to design protein structures with the desired catalytic activity. Since an effective enzyme catalysis depends on the stabilization of the transition state with the residues of the active site, various models of transition state (TS) such as small molecules mimicking the TS, optimized geometry of the TS in the gas phase, or modified optimized geometries are used to design synthetic and efficient enzymes.

Understanding the molecular basis of chiral discrimination in active sites is promising for biocatalysis and biotransformation. Once the interaction pattern in an active site is understood, then rational modification of the active site residues can alter the interaction between the active site and reactants in the desired manner and shall be able to control the reaction in a more efficient manner than the random trial-and-error approach. Once the influence of the pattern of chiral moieties in an active site is correlated with the possible reaction, the clues for constructing structural motifs as effective mimics of the archetypal biological active site can be obtained. The structure and functions of proteins are distinguished by certain structural motifs such as α-helix, β-sheet. Active sites and binding sites might be built up from amino acids, which are rather distant in their primary sequence. Such a distinct three-dimensional arrangement has an importance in functionality. The challenge is to preserve an active three-dimensional template observed in an enzyme in another molecular scaffold. The development of the minimal model of the template for targeted reactions with the aim of efficient miniaturization (over the large biological macromolecular machines) could be another arena of future development.

The design of bioactive compounds by incremental construction of a ligand model within a model of the receptor or enzyme active site, the structure of which is known from x-ray or NMR data (de novo design of active site), is an area of extensive research (Rao et al., 2007, 1320; Kim and Hecht, 2006, 15824; Hoang et al., 2006, 6883; Walter et al., 2005, 37742), Bradley et al., 2005, 201; Bilgicer and Kumar, 2004, 15324; Wei and Hecht, 2004, 67; Wei et al., 2003, 13270; Brown et al., 2003, 10712; Ellerby et al., 2003, 35311; Qamra et al., 2002, 967; Emberly et al., 2002, 11163; Wang and Hecht, 2002, 2760; Murphy et al., 2000, 149; Villain et al., 2000, 2676; West et al., 1999, 11211; Nedwidek and Hecht, 1997, 10010; Hellinga et al., 1997, 10015; Beasley and Hecht, 1997, 2031; Bryson et al., 1995, 935; Okada et al., 2000, 24630; Riechmann and Winter, 2000, 10068 and Xiang et al., 1999, 7638–7652). Protein design is carried out by creating binary patterns of nonpolar and polar (or hydrophobic and hydrophilic) residues where the sequence locations were specified explicitly, but the precise identities of the side chains were extensively varied. Characterization of the expressed proteins indicated that most of the designed sequences folded into compact alpha helical structures. It is suggested that a simple binary code of polar and nonpolar residues arranged in the appropriate order can drive polypeptide chains to collapse into a given structural motif (globular alpha-helical folds). (Kamtekar et al., 1993, 1680–1685). A number of studies are extended to develop strategies to design effective biomimetics.

It is suggested (Tatsis et al., 2008, 36) that use of procedures such as the mutation of a protein or by synthesizing short oligopeptides (less than 40 residues, which are termed *mini proteins*), it is possible to create a molecular environment capable of supporting the higher-level structural arrangement of the functional biological molecular fragments that are the hallmark of the protein. This procedure may reproduce the function of much larger proteins by transferring the active sites to small and stable natural scaffolds.

The oligopeptides are also developed for use as metal-free catalysts for enantioselective transformations of synthetic interest. Studies have been carried out on the screening of peptide libraries for ester and phosphate cleavage, aided by novel chromogenic and gel-based assays and optimization of metal-free peptide catalysts for asymmetric epoxidation, enantioselective acylation, and phosphorylation, conjugate addition to enimides, and hydrocyanation of imines (Strecker-reaction). Peptide scaffolds can also act as chiral binding sites, and can carry out enantioselective acyl transfer or a hydrocyanation reaction when nucleophilic catalyst *N*-methyl histidine or an imine-binding urea moiety (referred to as organocatalytic groups) are incorporated within the scaffold (Berkessel, 2003, 409). Mimics of biological active sites containing metals are also abundant in literature. The biologically important complexes containing zinc, cadmium, mercury and copper thiolates, and selenolates are prepared by self-assembly reactions starting from suitable precursor compounds, mostly in organic solvents. These molecules have the possibility to serve as functional mimics of protein active site features of biological archetypes such as metallothioneins (Henkel and Krebs, 2004, 801). However, it is at present not clear how the chirality of the primary level or secondary level of the molecular scaffold can be utilized to better improve the process of designing active site mimetics. It is necessary to understand the principles behind the role of chirality as present in biological active sites. These principles can then be systematically translated to create mimetic molecules to achieve strategies for effective asymmetric synthesis.

Considering the biotechnological activity and related utility of diverse natural products produced from organisms from all kingdoms of life, it has been long realized that pharmaceuticals can be developed by mimicking the related natural products. Interactions of chiral drugs with enzymes are important to investigate. Recent development of the detailed understanding of the drug–enzyme interaction such as with antibiotics provides impetus for such studies (Ramakrishnan, 2008, 567; Carter et al., 2000, 340; Brodersen et al., 2000, 1143; Brodersen et al., 2001, 17; Ramakrishnan, 2000). Combination of structural analysis based on crystallographic information and electronic structure-based computation should be a useful tool for such future developments. Recent significant progress in the quantum crystallography might have potential use in this direction (Huang et al., 1995, 371; Huang et al., 1996, 1691; Huang et al., 1999, 439; Huang et al., 2001, 409; Huang et al., 2005, 808; Huang et al., 2006a, 447; Huang et al., 2006b, 1233; Huang et al., 2007, 4261).

The development of the microscopic understanding of the chiral discrimination in the confined spaces of active sites has biomedical prospects. It is well known that the enzymes are highly efficient in synthesizing chiral drugs (Margolin, 1993, 266) and biocatalysis. A large proportion of clinically useful antibiotics exert their effects by blocking protein synthesis at PTC in ribosome (Hermann, 2005, 355). Understanding the recognition process at PTC has technological relevance for new drug design. This biomedical target is particularly important considering the problem of the recurrent development of antibiotic resistance. It is discussed in the present book that the modification of the structure of PTC can lead to alteration of chiral selectivity. Mutation of the PTC can lead to the achievement of controlled protein synthesis. More molecular studies are necessary to explore the recognition event in such cavities. A further point is that while the works described in the present book principally refer to the discrimination related to the thermodynamically stable state, the dynamic aspects of discrimination are also interesting. Confinement is already known to have an important influence on dynamics (Nandi et al., 2000: 2013). The dynamics of the biological active sites and its influence on the confined chiral molecules would be interesting to explore and might reveal new aspects of chiral recognition in biology and medicine. A study of the mechanism of action of chiral pollutants is another important future field of development where similar microscopic analysis could be useful (Ali and Aboul-Enein, 2004).

Enzymes are most efficient biocatalysts that carries out natural product metabolism reactions of diverse types. Enzyme engineering strategies have allowed the exploration of metabolic engineering of biosynthetic pathways to create new products that have biotechnological and medicinal implications. A basic understanding of the activity of these enzymes can be useful to obtain novel biocatalysts for industrial biotransformations (Bernhardt and O'Connor, 2009: 35). Biocatalysis is also a major current standard technology for the production of chemicals of which major products are chiral (89%) and are used as fine chemicals (Straathof et al., 2002: 548). Fine chemicals are pure, single chemical substances that are commercially produced with chemical reactions into highly specialized applications. Examples are active pharmaceutical ingredients and their intermediates, biocides (pesticides, herbicides, and other specialized chemicals that are used in agriculture to inhibit or kill pests and weeds and thus improve crop yields), and chemicals for technical applications (inks, performance-enhancing additives, special coatings, and photographic chemicals). An analysis of 134 industrial biotransformations reveals that hydrolases (44%) and redox biocatalysts (30%) are the most prominent categories. For the enantiomerically pure products, the chiral configuration originates directly from the precursors. Where the enantiomeric purity is due to the biotransformation, both kinetic resolution and asymmetric synthesis frequently occur. Asymmetric syntheses are mostly carried out by oxidoreductases and lyases, whereas kinetic resolutions almost exclusively involve hydrolases. In about a quarter of these kinetic resolutions, the hydrolases are applied in the synthetic rather than in the hydrolytic mode. The sources of chirality in the industrial biotransformations are enantiopure precursor (~42.5%), kinetic resolution (~25.6%), and asymmetric synthesis (21.1%). No chirality is involved in ~10.8% transformations (Straathof, 2002, 548). Out of the enzymes used in biotransformation, maximum used are hydrolases (~44.30%) oxidizing cells (~19.9%), lyases (~11.5%), transferase (~11.1%), reducing cells (~6.9%), isomerase (~3.5%) and oxidoreductase (~2.8%) (Straathof, 2002, 548). Analyses of the interactions in the respective active sites are necessary to develop new biocatalysts.

New catalytic principles leading to the efficient enantioselective synthesis of chiral nonracemic compounds are topics of extensive research (Turner, 2003, 401). Alternatives to asymmetric synthesis or kinetic resolution include dynamic kinetic resolution, and deracemisation and enantioconvergent transformations. Importantly, the parameters that can affect the stereochemical outcome of reactions (e.g., solvent, substrate design, immobilization, and directed evolution) are being better understood. As mentioned in the introduction, that the immobilization or confinement of molecules can enhance chiral discrimination is now a well-understood principle from computational studies in biomimetic systems. Despite the inherent complexity of enzyme-catalyzed reactions compared to those involving low-molecular-weight nonproteinogenic chiral catalysts, understanding various factors that can influence the stereochemical outcome of the reaction is difficult. In some cases, the effects of varying certain parameters (e.g., solvent, pH, and immobilization support) can be dramatic. Due to the increased availability of diverse genomic libraries, together with directed evolution technologies, the ability to tailor enzyme specificity in the desired direction is increasing substantially.

Röthlisberger and co-workers used computational techniques to design eight enzymes that use two different catalytic motifs to catalyze the Kemp elimination reaction with a measured rate enhancement of up to 10^5 and multiple turnovers. Choosing a particular catalytic mechanism, the authors used quantum mechanical transition state calculation to create an idealized active site with protein functional groups positioned so as to maximize transition state stabilization. Additional functional groups are included in the idealized active sites for two different mechanisms (carboxylate and histidine based) to further facilitate the catalytic activity. Density functional theory (B3LYP/6-31G(d)) is applied to optimize the placement and orientation of the catalytic groups around the transition state for maximal stabilization. To stabilize the transition state by charge delocalization, aromatic amino acid side chains are stacked above and below the planar transition state using effective π-stacking geometry (Röthlisberger et al., 2008, 190). Using mutational analysis, the authors confirmed that catalysis depends on computationally designed active site. X-ray structures are used to determine the geometries. An increase in catalytic activity as well as in turnover was achieved. Other examples of the utilization of computational methodology in designing biocatalysts are available in the literature (Jiang et al., 2008, 1387). The authors designed retro-aldolases that use four different catalytic motifs to catalyze the breaking of a carbon–carbon bond in a nonnatural substrate. The study demonstrates a computation based redesign of a novel retroaldolase activity in a scaffold that catalyzes an unrelated reaction.

These studies confirm that understanding the interaction of the active site residues is essential, as their mutual distance and orientation around the reactants significantly influence the course of the reaction and the stereoselectivity. Hahn and co-workers designed and synthesized a peptide having chymotrypsin-like esterase activity (Hahn et al., 1990, 1544). The designed peptide is a bundle of four short parallel helical peptides, linked covalently at their carboxyl ends and including the serine protease catalytic site residues serine, histidine, and aspartic acid. The spatial arrangement of the residues is kept the same as in chymotrypsin. The orientation-dependent interaction of the essential active site residues with the reactants that are vital for enzymatic activity is thereby retained in the synthetic system. The necessary oxyanion hole and substrate-binding pocket for acetyltyrosine ethyl ester were also included in the design. The peptide has affinity for chymotrypsin ester substrates similar to that of chymotrypsin, and hydrolyzes them at rates 1/100 slower than that of chymotrypsin. Total turnovers are greater than 100. The calculation of free energy profiles of the reaction from molecular dynamics simulation using a hybrid QM/MM system is another method to design efficient and selective biological catalysts, starting from the existing catalysts. The method allows detection of all specific substrate–protein interactions that are present in the active site of a catalyst.

Active site analogues of cytochrome P450, chloroperoxidase (CPO), and β-carotene monooxygenase are prepared using metalloporphyrine structures (Woggon, 2005, 127). The catalytically active systems exhibit chemical reactivity, which is comparable to native proteins. The limitations in reducing the protein structure to its center of chemical reactivity are discussed.

The foregoing limitations of the minimal model of the active site can partly be improved by incorporating second sphere residues that are involved in the reactions. This is shown in the case of Fe- and Mn-dependent superoxide dismutases using spectroscopic and computational studies (Jackson and Brunold, 2004, 461). Several second sphere residues are coupled with the hydrogen bond network and effectively discriminate substrate and hydroxyl ion, orienting incoming substrate for rapid electron transfer with the active site metal ion. It is possible that the incorporation of second sphere residues into the active site analogues might be useful in achieving better stereo-selectivity and fidelity. It is interesting to note that a 320 residue fragment of homodimeric *E. coli* HisRS is shown to catalyze both specific aminoacylation of tRNA and pyrophosphate exchange (Augustine and Francklyn, 1997, 3473). The efficiency is less than the full-length enzyme. Despite the limitations, such minimalist models of enzyme structures and comparison of their interactions with the full-length enzyme will help in designing new and efficient biocatalysts.

Chiral selectivity exhibited by ribozymes and DNA polymerases, which are related to the active site structure within RNA or DNA in relation to transferases, are discussed in this book. Although the molecular chirality is not introduced by RNA or DNA, the results indicate the chiral discrimination capabilities of RNA and DNA structures. The chirality is inherent in DNA and RNA structures at their various levels of structural hierarchy. The possible use of these chiral structures to induce enantioselectivity in chemical synthesis has already been pointed out in recent reviews (Roelfes, 2007, 126). Chirality transfer by nucleotide templated synthesis, enantioselective catalysis by RNA or DNA enzymes and DNA-based asymmetric catalysis are strategies employed for utilizing the chiral selectivity exhibited by these structures toward the development of scaffolds for new bioinspired catalytic systems.

The lessons learned from the active site structural features can be utilized in making the process of enzyme engineering more effective. If the principles of stereoselectivity and fidelity of the enzyme catalysis are understood, then it will be possible to develop new enzymes using either minimal modification of natural enzymatic structure or developing de novo scaffolds. An understanding of the conserved structural scaffolds, particularly near the reaction center in the active site, is important in this regard. Closer mutations near the substraate binding site or active site are better in many cases (Morley and Kazlauskas, 2005, 231–237). Efficient preparation of template enzymes and identification of the most suitable candidates are two important steps in enzyme engineering. Preparation of the desired enzyme can be achieved either by random trial and error (a stochastic method in which the random changes of amino acids are made without caring about their position or function) or through rational design (Otten et al., 2010, 46–54). A part of the rational design is the combination of crystal structure analysis and modeling. Various methods such as random methods, semirational design, or quantitative structure–activity relationship studies are also applied to the design of effective enzymes. It has been pointed out that the decision to choose a certain amino acid often is based on guesswork rather than a rational decision. The analysis of the interaction pattern in the active site can be useful in making the tailored enzyme.

While the foregoing conclusions provide a molecular understanding of the retention of biological homochirality and the related questions raised in the introduction, the understanding might be useful in following the principles of chiral molecular recognition in diverse systems. It is possible to extend the principles toward the material systems in order to gain control over the technique of nanofabrication of designer materials. Specific structures using chiral subunits (or achiral subunits in some cases) can be developed, which might have improved function. Here, we mention a few studies where the influence of the chirality and discrimination may be noted while building diverse nanomaterials. Future studies of the exploration of the role of molecular chirality and confinement might be interesting and useful.

We discussed some of the issues concerning the principles behind Nature's technique in building functional biological molecules, retaining their chiral structure, and replicatng these structures through evolution, as well as the accurate performance of the corresponding life processes. Combining these principles with the technological advancement already made might lead to the possible development of new synthetic architecture with the desired functionality. The basic units of known materials of technological interest such as diverse polymers, crystals, liquid crystals, and gels are rather simple in structure. Biological functional architectures are much more sophisticated. Nature has always been superior in designing the world through molecules on which the very existence of life depends. The immensely efficient and orchestrated synthesis of myriad biomolecules uses biophysical principles to achieve this. This book indicates that the confinement of biomolecules in a nanoscale dimension and their chirality is utilized in biosynthetic reactions to enhance the fidelity. Thus, gaining control over the methods of confinement of chiral moieties in a nanospace has vast potential in building up newer functional structures with desired physicochemical characteristics.

The principles learned can be extended to tunable synthesis of useful materials. Following, we mention a few possibilities of using the knowledge about the chiral discrimination in active sites to build usable materials. Such principles are already known for biomimetic systems (Ariga et al., 2008, 23; Ariga et al., 2006, 465). Numerous nanomaterials are designed through the process of self-assembly and molecular recognition, and specifically chiral recognition is important here (Ariga et al., 2008, 23; Ariga et al., 2006, 465; Ariga and Kunitake, 1998, 371; Kunitake, 1996, 9; Sakurai et al., 1997a, 4810; Sakurai et al., 1997b, 4817). In a recent study, excellent control of enantioselectivity of amino acid recognition is achieved by cholesterol-armed cyclen monolayer at the air–water interface (Michinobu et al., 2006, 14478). The host system is composed of cholesterol, which is chiral, and the Na^+ complex of cyclen has helical structure. Consequently, the system has multiple levels of chirality and exhibits significant enantioselectivity with leucine and valine. When leucine or valine was added to the subphase, the isotherms were shifted to larger molecular areas with increasing amino acid concentration. The isotherms in the presence of aqueous D-leucine are shifted to larger molecular areas than the same in the presence of aqueous L-leucine of the same concentration exhibiting discrimination. In contrast, the difference between the isotherms on aqueous L- and D-valines is insignificant. While these results suggest a greater and negligible enantioselectivity for leucine and valine, significant chiral discrimination is noted also for valine when the binding constant (K) is studied. The K values of D-leucine are always greater than those of

L-leucine, indicating that the monolayers have a stronger interaction with D-leucine. Conversely, the values of L-valine are smaller than those of D-valine at low surface pressure but exceed them at 22–23 mN m^{-1}. The result shows that the chiral discrimination and concomitant recognition in monolayers with valine changes from the D- to L-form upon compression. The result shows for the first time that the subtle difference in the chemical structure between leucine and valine can be distinguished by the dynamic process of monolayer formation. The multilevel chirality of the host, dissymmetry of the air–water interface, and the confinement in the nanoscale proximity of aggregate seem to have substantial influence on the recognition observed. The result shows that it is possible to construct nanoscale assemblies by which one can gain better control of enantioselectivity.

It is discussed (in the context of peptidyl transferase) that when confined (with restricted degrees of freedom), achiral molecular segments such as nucleobases can influence the discrimination in biological systems. It is noted in biomimetic models that achiral shapes arranged in a chiral (helical) way might have dissymmetric interaction profile with a ligand (a signature of chiral discrimination) (Nandi, 2004, 789). Synthetic molecule 2-(heptadecyl) naphtha-[2,3]-imidazole was found to form a bilayer film on a water surface. The in situ coordinated LB films composed of the molecule and Ag$^+$ ion showed chirality due to the formation of a helical coordination polymer where the naphtha-[2,3]-imidazole group distorted in a certain degree along the polymer backbone. Notably, both ligand and metal ions are achiral (Yuan and Liu, 2003, 5051). The coordination and the interaction between the naphtha [2,3] imidazole rings play an important role in forming the chiral LB films. The results hint at the possibilities of the design and fabrication of chiral molecular assemblies from achiral molecules.

The importance of the controlled fabrication of nanometer-scale objects, which is a central issue in current science and technology, has been recently discussed in a detailed review (Ariga et al., 2009, 014109). Diverse materials are designed by self-assembly, which have technical utility or might have potential in that direction. In a recent study, inorganic nanorods are used to develop optically driven nanorotors. The principle of this nanorotor is dependent on the nonzero chirality of the nanorod structure (Khan et al., 2006, S287). The inherent textural irregularities or morphological asymmetries of the nanorods give rise to the torque under the radiation pressure. The presence of nonzero chirality (arising from a small surface irregularity or extrusion) is sufficient to produce enough torque for reasonable rotational speed. As the extrusion is not symmetric about a plane passing through the z-axis, it will provide a net chirality of the nanorod structure. Therefore, the radiation pressure force will generate a nonzero torque on the nanorod. The controllability, simplicity, and flexibility of the rotational motion of these rotors suggest that this scheme can be applied universally to devise easy-to-use optically driven nano-rotors. The availability of inorganic nanorods bearing the desired size and chirality would favor the present scheme to design nanorotors with predictable rotational speeds. It might be possible that nanostructures and nanomachines with controllability can be constructed by tuning the chirality of the constituent nanomaterial. Exploiting nanospaces by the confinement of the immobilized chiral catalysts is found to be useful in asymmetric catalysis, which is an extremely important field of research in general (Thomas and Raja, 2008, 708).

The study of the role of the confined chiral moieties in the process of biological catalysis and the importance of the effect of the spatial restrictions imposed by the catalytic systems might be useful to explore the development of novel catalysts with enhanced function. A variety of complex biological and biomimetic systems have microheterogene-ity. The environment of the nanospace of the active site as discussed in the present book is one of them. Detailed studies have revealed that enhanced molecular interaction and multisite interaction are also fea-tures of the recognition phenomena at the microscopically heteroge-neous interfaces (Ariga and Kunitake, 31, 371; Kunitake, 1996, 9; Sakurai et al., 1997a, 4810; Sakurai et al., 1997b, 4817). The discriminations exhibited by the enzymes are essentially controlled by the preferential interactions at multiple sites. It is worthwhile to explore the unifying principles underlying the preferential interaction and discrimination in biological systems.

References

Ali, I. and Aboul-Enein, H.Y. (2004). Chiral pollutants: Sources and distribution. In *Chiral Pollutants: Distribution, Toxicity and Analysis by Chromatography*, Chichester, UK: John Wiley & Sons.

Ariga, K. and Kunitake, T. (1998). Molecular recognition at air-water and related interfaces: Complementary hydrogen bonding and multisite interaction. *Acc. Chem. Res.*, 31: 371–378.

Ariga, K., Hill, J.P., Lee, M.V., Vinu, A., Charvet, R., and Acharya, S. (2009). *Sci. Technol. Adv. Mater.* 9: 014109.

Ariga, K., Michinobu, T., Nakanishi, T., and Hill, J.P. (2008). Chiral recognition at the air–water interface. *Curr. Opin. Colloid Interface Sci.*, 13: 23–30.

Ariga, K., Nakanishi, T., and Hill, J.P. (2006). A paradigm shift in the field of molecular recognition at the air–water interface: From static to dynamic. *Soft Matter*, 2: 465–498.

Augustine, J. and Francklyn, C. (1997). Design of an active fragment of a class II aminoacyl tRNA synthetase and its significance for synthetase evolution. *Biochemistry*, 36: 3473–3482.

Beasley, J.R. and Hecht, M.H. (1997). Protein design: The choice of de novo sequences. *J. Biol. Chem.*, 272: 2031–2034.

Benner, S. (1993). Catalysis: Design versus selection. *Science*, 261: 1402–1403.

Benson, D.E., Wisz, M.S., and Hellinga, H.W. (2000). Rational design of nascent metalloenzymes. *Proc. Natl. Acad. Sci. (USA)*, 97: 6292–6297.

Berkessel, A. (2003). The discovery of catalytically active peptides through combinatorial chemistry. *Curr. Opin. Chem. Biol.*, 7: 409–419.

Bernhardt, P. and O'Connor, S.E. (2009). Opportunities for enzyme engineering in natural product biosynthesis. *Curr. Opin. Chem. Biol.*, 13: 35–42.

Bilgicer, B. and Kumar, K. (2004). De novo design of defined helical bundles in membrane environments. *Proc. Natl. Acad. Sci. (USA)*, 101: 15324–15329.

Bradley, L.H., Kleiner, R.E., Wang, A.F., Hecht, M.H., and Wood, D.W. (2005). An intein-based genetic selection allows the construction of a high-quality library of binary patterned de novo protein sequences. *Protein Eng. Des. Sel.*, 18: 201–207.

Brodersen, D.E., Carter, A.P., Clemons, W.M., Jr., Morgan-Warren, R.J., Murphy F.V. IV, Ogle, J.M., Tarry, M.J., Wimberly, B.T., and Ramakrishnan, V. (2001). Atomic structures of the 30S subunit and its complexes with ligands and antibiotics. *Cold Spring Harbor Symp. Quant. Biol.*, 66: 17–32.

Brodersen, D.E., Clemons, W.M., Jr., Carter, A.P., Morgan-Warren, R.J., Wimberly, B.T., and Ramakrishnan, V. (2000). The structural basis for the action of the antibiotics tetracycline, pactamycin and hygromycin B on the 30S ribosomal subunit. *Cell*, 103: 1143–1154.

Brown, S., Fawzi, N.J., and Head-Gordon, T. (2003). Coarse-grained sequences for protein folding and design. *Proc. Natl. Acad. Sci. (USA)*, 100: 10712–10717.

Bryson, J.W., Betz, S.F., Lu, H.S., Suich, D.J., Zhou, H.X., O'Neil, K.T., and DeGrado, W.F. (1995). Protein design: A hierarchic approach. *Science*, 270: 935–941.

Carter, A.P., Clemons, W.M., Jr., Brodersen, D.E., Morgan-Warren, R.J., Wimberly, B.T., and Ramakrishnan, V. (2000). Functional insights from the structure of the 30S ribosomal subunit and its interactions with antibiotics. *Nature*, 407: 340–348.

Ellerby, H.M., Lee, S., Ellerby, L.M., Chen, S., Kiyota, T., del Rio, G., Sugihara, G., Sun, Y., Bredesen, D.E., and Arap, W. (2003). An artificially designed pore-forming protein with anti-tumor effects. *J. Biol. Chem.*, 278: 35311–35316.

Emberly, E.G., Wingreen, N.S., and Tang, C. (2002). Designability of alpha-helical proteins. *Proc. Natl. Acad. Sci. (USA)*, 99: 11163–11168.

Fersht, A.R., Knill-Jones, J.W., Bedonelle, H., Winter, G. (1988). Reconstruction by site-directed mutagenesis of the formation state for the activation of tyrosine by the tyrosyl-tRNA synthetase: A mobile loop envelopes the transition state in an induced fit mechanism. *Biochemstry*, 27: 1581–1587.

Hahn, K.W., Klis, W.A., and Stewart, J.M. (1990). Design and synthesis of a peptide having chymotrypsin-like esterase activity. *Science*, 248: 1544–1547.

Handel, T., Williams, S., and DeGrado, W. (1993). Metal ion-dependent modulation of the dynamics of a designed protein. *Science*, 261: 879–885.

Hellinga, H.W. (1997). Rational protein design: Combining theory and experiment. *Proc. Natl. Acad. Sci. (USA)*, 94: 10015–10017.

Henkel, G. and Krebs, B. (2004). Metallothioneins: Zinc, cadmium, mercury, and copper thiolates and selenolates mimicking protein active site features—structural aspects and biological implications. *Chem. Rev.*, 104: 801–824.

Hermann, T. (2005). Drugs targeting the ribosome. *Curr. Opin. Struct. Biol.*, 15: 355–366.

Hoang, T.X., Marsella, L., Trovato, A., Seno, F. ,Banavar, J.R., and Maritan, A. (2006). Common attributes of native-state structures of proteins, disordered proteins, and amyloid. *Proc. Natl. Acad. Sci. (USA)*, 103: 6883–6888.

Horsman, G.P., Liu, A.M.F., Henke, E., Bornscheuer, U.T., and Kazlauskas, R.J. (2003). Mutations in distant residues moderately increase the enantioselectivity of *Pseudomonas fluorescens* esterase towards methyl 3-bromo-2-methyl-propanoate and ethyl 3-phenylbutyrate. *Chem. Eur. J.*, 9: 1933–1939.

Huang, L., Massa, L., and Karle, J. (1995). Quantum crystallography and the use of kernel projector matrices. *Int. J. Quantum Chem.: Quantum Chem. Symp.*, 29: 371–384.

Huang, L., Massa, L., and Karle, J. (1996). Kernel projector matrices for Leu^1-zervamicin. *Int. J. Quantum Chem.: Quantum Chem. Symp.* 30, 1691–1700.

Huang, L., Massa, L., and Karle, J. (1999). Quantum crystallography applied to crystalline maleic anhydride. *Int. J. Quantum Chem.*, 73: 439–450.

Huang, L., Massa, L., and Karle, J. (2001). Quantum crystallography, a developing area of computational chemistry extending to macromolecules. *J. Res.Dev.*, 45: 409–415.

Huang, L., Massa, L., and Karle, J. (2005). Kernel energy method illustrated with peptides. *Int. J. Quantum Chem.*, 103: 808–817.

Huang, L., Massa, L., and Karle, J. (2006a). Kernel energy method: Basis functions and quantum methods. *Int. J. Quantum Chem.*, 106: 447–457.

Huang, L., Massa, L., and Karle, J. (2006b). The kernel energy method: Application to a tRNA. *Proc. Natl. Acad. Sci. (USA)*, 103: 1233–1237.

Huang, L., Massa, L., and Karle, J. (2007). Drug target interaction energies by the kernel energy method in aminoglycoside drugs and ribosomal A site RNA targets. *Proc. Natl. Acad. Sci. (USA)*, 104: 4261–4266.

Jackson, T.A. and Brunold, T.C. (2004). Combined spectroscopic/computational studies on Fe and Mn dependent superoxide dismutases: Insights into second-sphere tuning of active site properties. *Acc. Chem. Res.*, 37: 461–470.

Jiang, L., Althoff, E.A., Clemente, F.R., Doyle, L., Röthlisberger, D., Zanghellini, A., Gallaher, J.L., Betker, J., Tanaka, F., Barbas, III, C.F., Hilvert, H., Houk, K.A., Stoddard, B.L., and Baker, D. (2008). De novo computational design of retro-aldol enzymes. *Science*, 319: 1387–1391.

Kamtekar, S., Schiffer, J.M., Xiong, H., Babik, J.M., and Hecht, M.H. (1993). Protein design by binary patterning of polar and nonpolar amino acids. *Science*, 262: 1680–1685.

Khan, M., Sood, A.K., Deepak, F.L., and Rao, C.N.R. (2006). Nanorotors using asymmetric inorganic nanorods in an optical trap. *Nanotechnology*, 17: S287–

Kim, W. and Hecht, M.H. (2006). Generic hydrophobic residues are sufficient to promote aggregation of the Alzheimer's Abeta42 peptide. *Proc. Natl. Acad. Sci. (USA)*, 103: 15824–15829.

Klis, W.A. and Stewart, J.M. (1990). Design and synthesis of a peptide having chymotrypsin-like esterase activity. *Science*, 248: 1544–1547.

Kunitake, T. (1996). Molecular recognition by molecular monolayers, bilayers, and films. *Thin Solid Films*, 284–285: 9–12.

Margolin, A.L. (1993). Enzymes in the synthesis of chiral drugs. *Enzyme Microb. Technol.*, 15: 266–280.

Marti, S., Andres, J., Moliner, V., Silla, E., Tunon, I., and Bertran, J. (2008). Computational design of biological catalysts, *Chem. Soc. Rev.*, 37: 2634–2643.

Michinobu, T., Shinoda, S., Nakanishi, T., Hill, J.P., Fuji, K., Player, T.N., Tsukube, H., and Ariga, K. (2006). Mechanical control of enantioselectivity of amino acid recognition by cholesterol-armed cyclen monolayer at the air-water interface. *J. Am. Chem. Soc.*, 128: 14478–14479.

Morley, K.L. and Kazlauskas, R.J. (2005). Improving enzyme properties: When are closer mutations better? *Trends Biotech.*, 23: 231–237.

Murphy, L.R., Wallqvist, A., and Levy, R.M. (2000). Simplified amino acid alphabets for protein fold recognition and implications for folding. *Protein Eng. Des. Sel.*, 13: 149–152.

Nandi, N. (2004). Role of secondary level chiral structure in the process of molecular recognition of ligand: Study of model helical peptide. *J. Phys. Chem. B.*, 108: 789–797.

Nandi, N., Bhattacharyya, K., and Bagchi, B. (2000). Dielectric relaxation and solvation dynamics of water in complex chemical and biological systems. *Chem. Rev.*, 100: 2013–2045.

Nedwidek, M.N. and Hecht, M.H. (1997). Minimized protein structures: A little goes a long way. *Proc. Natl. Acad. Sci. (USA)*, 94: 10010–10011.

Nureki, O., Vassylyev, D.G., Tateno, M., Shimada, A., Nakama, T., Fukai, S., Konno, M., Hendrickson, T.L., Schimmel, P., Yokoyama, S. (1998). Enzyme structure with two catalytic sites for double-sieve selection of substrate. *Science*, 280: 578–581.

Okada, Y., Ito, Y., Kikuchi, A., Nimura, Y., Yoshida, S., and Suzuki, M. (2000). Assignment of functional amino acids around the active site of human DNA topoisomerase II alpha. *J. Biol. Chem.*, 275: 24630–24638.

Otten, L.G., Hollmann, F., and Arends, I.W.C.E. (2009). Enzyme engineering for enantioselectivity: From trial-and-error to rational design? *Trends Biotech.*, 28: 46–54.

Oue, S., Okamoto, A., Yano, T., and Kagamiyama, H. (1999). Redesigning the substrate specificity of an enzyme by cumulative effects of the mutations of non-active site residues. *J. Biol. Chem.*, 274: 2344–2349.

Pérez-Payá, E., Houghten, R.A., and Blondelle, S.E. (1996). Functionalized protein-like structures from conformationally defined synthetic combinatorial libraries. *J. Biol. Chem.*, 271: 4120–4126.

Qamra, R., Taneja, B., and Mande, S.C. (2002). Identification of conserved residue patterns in small β-barrel proteins. *Protein Eng. Des. Sel.*, 15: 967–977.

Ramakrishnan, V. (2008). What we have learned from ribosome structures. *Biochem. Soc. Trans.*, 36: 567–574.

Rao, A., Ram, G., Saini, A.K., Vohra, R., Kumar, K., Singh, Y., and Ranganathan, A. (2007). Synthesis and selection of de novo proteins that bind and impede cellular functions of an essential mycobacterial protein. *Appl. Environ. Microbiol.*, 73: 1320–1331.

Riechmann, L. and Winter, G. (2000). Novel folded protein domains generated by combinatorial shuffling of polypeptide segments. *Proc. Natl. Acad. Sci. (USA)*, 97: 10068–10073.

Roelfes, G. (2007). DNA and RNA induced enantioselectivity in chemical synthesis. *Mol. BioSyst.*, 3: 126–135.

Röthlisberger, D., Khersonsky, O., Wollacott, A.M., Jiang, L., DeChancie, J., Betker, J., Gallaher, J.L., Althoff, E.A., Zanghellini, A., Dym, O., Albeck, S., Houk, K.A., Tawfik, D.S., and Baker, D. (2008). Kemp elimination catalysts by computational enzyme design. *Nature*, 453: 190–195.

Sakurai, M., Tamagawa, H., Inous, Y., Ariga, K., and Kunitake, T. (1997a). Theoretical study of intermolecular interaction at the lipid-water interface.1. Quantum chemical analysis using a reaction field theory. *J. Phys. Chem. B*, 101: 4810–4816.

Sakurai, M., Tamagawa, H., Inous, Y., Ariga, K., and Kunitake, T. (1997b). Theoretical study of intermolecular interaction at the lipid–water interface. 2. Analysis based on the Poisson–Boltzmann equation. *J. Phys. Chem. B*, 101: 4817–4825.

Straathof, A.J., Panke, S., and Schmid, A. (2002). The production of fine chemicals by biotransformations. *Curr. Opin. Biotechnol.*, 13: 548–556.

Tanner, M.E. (2002). Understanding Nature's strategies for enzyme-catalyzed racemization and epimerization. *Acc. Chem. Res.* 35: 237–246.

Tatsis, V.A., Stavrakoudis, A., and Demetropoulos, I.N. (2008). Molecular dynamics as a pattern recognition tool: An automated process detects peptides that preserve the 3D arrangement of trypsin's active site. *Biophys. Chem.*, 133: 36–44.

Thomas, J.M. and Raja, R. (2008). Exploiting nanospace for asymmetric catalysis: Confinement of immobilized, single-site chiral catalysts enhances enantioselectivity. *Acc. Chem. Res.*, 41: 708–720.

Turner, N.J. (2003). Controlling chirality. *Curr. Opin. Biotechnol.*, 14: 401–406.

Villain, M., Jackson, P.L., Manion, M.K., Dong, W.-J., Su, Z., Fassina, G., Johnson, T.M., Sakai, T.T., Krishna, N R , and Blalock, J.E. (2000). De novo design of peptides targeted to the EF hands of calmodulin. *J. Biol. Chem.*, 275: 2676–2685.

Walter, K.U., Vamvaca, K., and Hilvert, D. (2005). An active enzyme constructed from a 9-amino acid alphabet. *J. Biol. Chem.*, 280: 37742–37746.

Wang, W. and Hecht, M.H. (2002). Rationally designed mutations convert de novo amyloid-like fibrils into monomeric beta sheet proteins. *Proc Natl Acad. Sci. (USA)*, 99: 2760–2765.

Wei, Y., Kim, S., Fela, D., Baum, J., and Hecht, M.H. (2003). Solution structure of a de novo protein from a designed combinatorial library. *Proc. Natl. Acad. Sci. (USA)*, 100: 13270–13273.

Wei, Y. and Hecht, M.H. (2004). Enzyme-like proteins from an unselected library of designed amino acid sequences. *Protein Eng. Des. Sel.*, 17: 67–75.

West, M.W., Wang, W., Patterson, J., Mancias, J.D., Beasley, J.R., and Hecht, M.H. (1999). De novo amyloid proteins from designed combinatorial libraries. *Proc. Natl. Acad. Sci. (USA)*, 96: 11211–11216.

Woggon, W.-D. (2005). Metalloporphyrines as active site analogues—lessons from enzymes and enzyme models. *Acc. Chem. Res.*, 38: 127–136.

Xiang, H., Luo, L., Taylor, K.L., and Dunaway-Mariano, D. (1999). Interchange of catalytic activity within the 2-enoyl-coenzyme A hydratase/isomerase superfamily based on a common active site template. *Biochemistry*, 38: 7638–7652.

Yuan, J. and Liu, M. (2003). Chiral molecular assemblies from a novel achiral amphiphilic 2-(heptadecyl) naphtha[2,3]imidazole through interfacial coordination. *J. Am. Chem. Soc.*, 125: 5051–5056.

Index